Diesel & Electric Locomotives Around The UK
IN THE EIGHTIES & NINETIES

Roger Rounce

Published in Great Britain by Gresley Books
An imprint of Mortons Books Ltd.
Media Centre
Morton Way
Horncastle LN9 6JR
www.mortonsbooks.co.uk

© 2024 by Gresley Books

All rights reserved. No part of this publication may be reproduced or transmitted in any form or by any means, electronic or mechanical including photocopying, recording, or any information storage retrieval system without prior permission in writing from the publisher.

ISBN 978-1-911704-22-5

The right of Roger Rounce to be identified as the author of this work has been asserted in accordance with the Copyright, Designs and Patents Act 1988.
Typeset by Jayne Clements (jayne@hinoki.co.uk), Hinoki Design and Typesetting

Acknowledgements
My thanks are due, once again, to Michael Fanthorpe, a friend of many years and someone with whom I have shared so many railway adventures. When occasionally my own notes appear incomplete, I can always rely on his expertise, attention to detail and methodical recording of times and places. I must also thank Tim Mercer, a very knowledgeable railway enthusiast and modeller, Keith Renaut for his 'Underground' expertise, Richard Garrod for his help in the motor vehicle department and Roger Harris publications.

Thanks
I imagine that younger readers wonder how enthusiasts in The Eighties and Nineties were able to wander around most depots, seemingly at will, making notes and taking photographs. This is because it was a different time, then, before political correctness had gone mad and when health and safety (if there was such a thing, then) adopted a sensible, measured approach. So, thanks are surely due to every foreman and every member of depot staff who turned a blind eye to all of us enthusiasts pursuing our fascinating yet harmless hobby. Without them, this book and many others like it, would not exist.

Front cover: Class 56 56111 at Knottingley TMD on 27 April 1991.
Title page: Built by English Electric in 1965, Class 86/2 86228 (WN) Vulcan Heritage (E3167) sits at Liverpool Street station on 19 March 1987 in InterCity livery; these stabling roads went with subsequent modernisation, along with the cabin.
Back cover: Newly delivered Class 60 60094 Tryfan at Leicester TMD on 16 February 1992.

Introduction

As in Stratford Depot Locomotives in The Eighties and Nineties, the running number, shed code and name (if any) of each locomotive at the time of the photograph are shown in bold type; previous numbers carried are bracketed following it and any numbers it was to receive in the future are bracketed preceding it. This is illustrated in the following example: (47776) **47578 (ED) The Royal Society of Edinburgh** (47181) (D1776). The only exception to this is the section regarding the last days of **'The European'** where I thought it appropriate to highlight the train name in bold type whenever it is referred to, albeit not in italics. During the period covered by both books, I was employed by London Clubs Ltd. in the Cage (Cashdesk) at a West End casino; depending upon my shift, I travelled to London each day by car or train and found that the nature of the hours and shifts were conducive to the pursuance of other interests, including railways.

This is a 'round the UK tour' that will hopefully appeal to enthusiasts from Scotland to Devon and Cornwall. As you can imagine, I had so many images to choose from (over 400 at Euston and Kings Cross alone) and therefore selecting what to include was difficult, but I hope I have picked out a few from each location that fall into either the unusual or interesting bracket.

This book is intended to be both informative and entertaining, with images of diesel and electric locomotives that many of us are lucky enough to remember seeing at the time and many will wish they had; hopefully, the locomotive names, liveries and locations will bring back many fond memories. These days, the default of many enthusiasts is the internet but this is an ephemeral medium, gone like early morning mist when you turn off the computer; nothing beats a book. Dip into it; there is a lot here and I hope the reader is constantly finding something new to read or view for quite a while.

Comments and observations are welcome using my email mightyiron@hotmail.com or my website www.rogerrounce.com

KNOTTINGLEY TMD

This general view of Knottingley TMD was to be had from the 'grassy knoll', as we called it, a cluster of trees at the top of the slope that led down into the Depot; the shed staff allowed the trees to grow out of hand to deter enthusiasts from spotting locos. The faint-hearted never ventured further than there but I usually drove in; it seemed easier and, after all, locos inside the shed could not be seen from that position.

Left: On 20 May 1989, Doncaster-built Class 56 (69014) **56104 (TO)** sits in the spring sunshine at Knottingley Depot unaware what the future held in store for it. Because, in 2022, GB Railfreight donated the power unit, alternator and valuable spares from 56104 to the Ivatt Diesel Recreation Society for use on LMS 10000. It was delivered to the Ecclesbourne Valley Railway and was intended to be a back-up to the English Electric power unit they already had to speed up work on the new 'Class 69' at Progress Rail. The chassis for this new enterprise came from 58022. Going new to Sheffield (Tinsley) in 1982, during its life, 56104 also worked out of Toton, Motherwell and Immingham. It carries Railfreight triple-grey 'Coal' livery. The Railfreight director at the time said the black diamonds depicted 'knobs of coal'. Hmm. However, after donating most of its innards to LMS 10000, it was then chosen to be reconstructed as a Class 69.

Right: Built by Brush Traction, **60099 (TO)** *Ben More Assynt* went new to Toton in 1992, later allocations being Stewarts Lane and Thornaby. It is carrying 'Mainline' livery on 2 August 1997 and, by this date, '60s' and '66s' were beginning to make inroads into the '56s' and '58s' at Knottingley. A world away from my first visits when '47s' were the order of the day. Along with many others of the Class, 60099 was held in 'Tactical Reserve' store at Toton but is now operational again.

Left: Derby-built in 1958, **09201** (08421) (D3536) was a Scottish loco for the first 30 years of its life. New to Stirling, further allocations were Eastfield (Glasgow), Dundee, Motherwell and Haymarket (Edinburgh). It then worked out of Gateshead, Blyth, Knottingley, Toton and Doncaster, arriving at Knottingley for a second time in 1998. 09201 was taken by road to Thornaby for repairs in 2006 and was later stored at Toton and Crewe Electric. Sold to Harry Needle in 2015, it was moved to Hope Cement Works and thence to Barrow Hill for storage. This image is dated 26 April 1998.

Right: Doncaster-built **56033 (IM)** *Shotton Paper Mill* went new to Toton in August 1977 but was badly damaged in an accident three months later and went back to Doncaster Works for repair. It later worked out of Sheffield (Tinsley), Cardiff Canton, Healey Mills, Bristol (Bath Road) and Stewarts Lane. It was broken up by European Metal Recycling, Kingsbury, in 2010. The shunter is 08597 (KY); there was always a good turnover over of '08s' at Knottingley, and this one stayed a couple of years. The date is 14 February 1998.

Built by Brush in 1959, Class 31/1 **31146 (BS)** *Brush Veteran* (D5564) began its working life at March. Further allocations were Norwich (Thorpe), York, Leeds (Holbeck), Sheffield (Tinsley), Immingham, Healey Mills, Crewe Diesel and Bescot, where it was named Brush Veteran in 1992. It is displaying 'Dutch' livery on 5 July 1998 and a few months later it would suffer a main generator fire and be withdrawn from service. 31146 was cut up by Booth Roe Metals, Rotherham, in 2002. I also photographed it in standard large logo Railfreight Grey when 'nameless' in March TMD on 11 April 1987.

Right: Doncaster-built Class 58 **58009** refuels whilst obeying the red flag at the same time. It went new to Toton in 1983 where it stayed until reallocated to Eastleigh in 2001. Given 'Special Projects' status, it made two trips to France via the Channel Tunnel and, the second time, stayed there. It has been in store at Alizay since 2016. Not at the palace, I assume. This image is dated 2 August 1997.

Left: A misty morning at Knottingley and one of the less-photographed Class 60s is the first loco I see. Now sporting 'Mainline' livery, Brush Traction-built Class 60 **60075 (TO)** *Liathach* went new to Toton in 1975. It is currently held in 'Tactical Reserve' store at Toton. The date is 21 December 1997.

Doncaster-built Class 56 **56111 (TO)** went new to Healey Mills in 1982 but was soon back at Doncaster with vibration problems and stayed there nearly three months so they were obviously not 'Good Vibrations'. Later allocations were Sheffield (Tinsley), Toton**,** Immingham and Thornaby, where it went into storage and from where it was eventually withdrawn. It was moved by road to European Metal Recycling, Kingsbury, and broken up in 2011. Twenty years prior to that, I photographed it at Knottingley on 27 April 1991.

Left: Cop! The first shall be last! I remember this moment very well because the first, was indeed, the last. The first Class 37 that I had ever seen at Knottingley was also the last I needed to complete my sighting of every loco in that Class. The date was 12 May 1989 and I find it hard to believe it was over 30 years ago. English Electric-built Class 37/0 **37212 (CF)** (D6912) in standard Rail Blue was a Cardiff Canton loco at the time as the shedcode sticker correctly indicates; beginning life in 1964 at Landore (Swansea), it moved on to Cardiff Canton, Sheffield (Tinsley), Gateshead, Thornaby, Eastfield (Glasgow), Motherwell, Inverness, Bescot, Crewe Diesel, Immingham, Toton and Eastleigh. Withdrawn in 1999, it was cut up at Eastleigh in 2004.

Right: Doncaster-built Class 58 **58040 (TO)** *Cottam Power Station* went new to Toton in 1986. It was stored unserviceable at Stratford in 1999, staying for two years before being towed to Ipswich Wagon repair Depot. Apparently, the fault wasn't terminal because it went to France in 2008, returning in 2010 to Eastleigh where it was put into 'Tactical Store'. Not for long, however, because the next year it was to be found back in France, in store at Alizay and was still there in 2016. This image of 58040 in 'Mainline' livery appears to be a case of 'one coat Magicote' as the old advertisement used to assure us because Red Stripe livery is showing through already. The date is 28 February 1998.

Left: Strips of webbed rubber matting were used in diesel depots to stop railwaymen slipping on the inevitable patches and puddles of diesel oil; I found it quite useful, too, and an example is seen here next to Doncaster-built **56110 (TO)** *Croft*, carrying the attractive 'Loadhaul' livery. New to Healey Mills in 1982, future allocations were Sheffield (Tinsley), Toton, Immingham and Thornaby. Stored for two years from 2008, it was taken by road to European Metal Recycling, Attercliffe, and broken up in 2011. The date is 1 September 1995.

Right: New to Toton in 1984, Doncaster-built **58021 (TO)** *Hither Green Depot*. It was reallocated to Eastleigh in 2001 and was sent through the Channel Tunnel to France in 2009. Stored at Alizay, it was still there in 2016. This image of it in 'Mainline' livery is dated 5 July 1998. Hither Green chose Oast Houses to be depicted on nameplates and Depot Plaques; apparently, they were supposed to represent Kent which, as Hither Green is in London, never quite made sense to me.

Above: Gaining access to the National Power Depot at Ferrybridge was extremely difficult and I know of nobody else among my peers who achieved this feat; Class 59/2 **59201 (NP)** *Vale of York* had only been in service for three months when I managed to photograph it at its home base. The livery was a bright blue main body over a thin pale grey strip with narrow white and red stripes running along the side between the blue and grey. Signal yellow was applied below the cab windows and a white and red National Power logo was positioned in the centre of the sides. In 1998, National Power ceased operating their own trains and the fleet was sold to EWS (English, Welsh and Scottish Railway) who deployed them on stone trains alongside Mendip Rail's 59/0s and 59/1s. The date is 22 June 1994; it is still operational.

Below: At the same time, I photographed this privately owned 0-4-0 DE Ruston and Hornsby ex-Staythorpe Power Station loco which displayed **'No. 1B'**. It was ex-Works number 420137 and was scrapped by C. F. Booth of Rotherham in 2009.

Above: Six years later on 24 July 1999, **59201 (TO)** *Vale of York* sits at Knottingley TMD in EWS colours. Strange how we enthusiasts move heaven and earth to see locos we need, only to find that they become very common years later.

Right: Moving on in the life of **59201 (MD)**, the date of the inset image is 27 July 2016 and, until March of that year, it had been owned by DB Schenker when the company was rebranded DB Cargo Ltd. I went to see 59201 at Ferrybridge in 1994 and now it has come to Chelmsford to see me. The ex-GE main line runs above these aggregate sidings and, as my son lives close by, I see quite a few of the comings and goings of various classes of diesels here.

BOUNDS GREEN TMD

Right: On 16 September 1989, Derby-built Class 08 **08222 (WDN)** (D3292) (13292) is stored out of use in the long grass at the back of Bounds Green TMD. It had been withdrawn in 1984 and remained in ever more deteriorating condition until, in 1997, it was moved to Old Oak Common and cut up on site by M. R. J. Phillips of Llanelly. Going new to Sheffield (Brightside/Grimesthorpe) in 1956, it also worked out of Toton, Springs Branch (Wigan) and Crewe Diesel, moving to Bounds Green in 1983.

Left: Carrying standard large logo Railfreight Grey livery, Brush-built **47348 (IM)** *St. Christopher's Railway Home* (D1829) first worked in a division of the Midland Region when new in 1964 until reallocated to Toton in 1973; Crewe Diesel came next, and it gained its name at Coalville Open Day in 1987. Further allocations were Immingham, Bristol (Bath Road), Stratford, Sheffield (Tinsley), Bescot and back to Crewe Diesel in 1999, specifically to move the new 'Juniper' EMUs from Alstom Works, Birmingham, to various locations. Whilst performing an EMU stock move to Shoeburyness in 2000 it became damaged when derailed and was put into store at Stratford. Years of storage followed until 47348 was eventually cut up in 2007 by Ron Hull Junior, Rotherham. Is it me, or do those wires look a bit too close to 47348's roof for comfort? I note that I saw this loco in Bow Yard when it was shedded at Immingham on 12 January 1989; this image is dated 16 September 1989.

St Christopher's Railway Home was established in January 1875 as the Railway Servants' Orphanage. It was intended for the children of railway workers who had lost their lives in the performance of their duty but, from 1881, the children of railway workers who had died of natural causes were also accepted and, from 1927, those whose mothers had died or were incapacitated. Children whose health was poor were not admitted but were granted allowances for their maintenance at home or at a special school.

Roll out the barrels; a rare example of a TOPS locomotive running-number ending in '00'. Class 31/4 **31400 (CD)** (31161) (D5579) sits at Bounds Green on 16 September 1989, a wet morning. It is hard to believe that this is the loco I had photographed at Stratford Depot 25 years previously in experimental bronze gold livery (see inset). Beginning service at Stratford in 1960, future sheds were March, Sheffield (Tinsley), Immingham, Finsbury Park and Norwich (Thorpe) until withdrawal in 1988 due to fire damage. After reinstatement, it worked out of Crewe Diesel and Bescot until it became involved in a collision with 31514 at St. Pancras station in January 1991. Although repaired, it was soon the subject of storage and was withdrawn for the last time in July 1991. 31400 was cut up by Booth Roe, Rotherham, but not until 1993.

BRUSH 'FALCON' WORKS LOUGHBOROUGH

These images were taken from Loughborough Meadows, through a chain-link fence a few hundred yards from the Works. However, it is surely interesting to see Class 60s newly built and sitting in the yard at Brush, albeit behind wire fencing. Getting to the Meadows wasn't exactly a piece of cake, either; access was gained by scrambling over a fence close to a railway bridge on the A60 Nottingham Road, climbing to the top of the embankment, and walking along the trackbed of the ex-Great Central Railway. I saw most of the new 60s at Loughborough but a few escaped me; however; I saw them all soon enough when they were working. I remember thinking that it seemed strange to be collecting numbers that I had always associated with A4 and A3 Pacific steam locomotives.

Left: On 9 June 1990, **60022 Ingleborough**, **60021 Pen-y-Ghent**, and **60020 Great Whernside**, will soon be on their way to their first Depots.

Right: On 26 January 1991, likewise for **60039 Glastonbury Tor** and **60045 Josephine Butler**.

Above: On 19 July 1991, **60069** *Humphry Davy* is clearly new but **60011** *Cader Idris* looks as though it has returned for the rectification of a fault.

Right: On the same date as above, **60073** *Cairn Gorm* (left) and **60074** *Braeriach* (right) appear to be anonymous but, as the numbers were written in my book on the day, I can only assume they appeared on the other ends. I have shown the names they would be receiving any day now.

AYR TMD

This was my first visit to Ayr since steam days and there were 13 diesels on shed, meagre fare in comparison to a shed full of 'Crabs' and 'Black 5s'. However, on 22 August 1989, it was good to photograph Class 26/0 **26004 (ED)** (D5304) which I note I saw in both Hornsey shed and Kings Cross station when new in 1958. Of course, that was when it carried the original livery that showed this class off to its best; British Railways green, white cab window surrounds, grey roof and a thin white stripe midway up the bodysides. After Hornsey, allocations were Finsbury Park, Haymarket (Edinburgh) and Eastfield (Glasgow). Withdrawn in 1992, this loco is now preserved. Here, it is carrying Railfreight triple grey livery and an Eastfield 'Scotty Dog' (sorry, West Highland Terrier) Depot plaque.

Class 26/1 **26026 (ED)** (D5326) was not so lucky, being cut up by M. C. Metal Processing, Springburn Works, Glasgow, in 1995. Prior to this, it had worked out of Haymarket (Edinburgh), Inverness and Eastfield (Glasgow). New in 1959, like 26004, it had been built by the Birmingham Railway Carriage & Wagon Company. This image of 26026 carrying an Eastfield 'Scotty Dog' logo in addition to the Railfreight General Grey livery is also dated 22 August 1989.

BASINGSTOKE STATION

Left: Class 50 **50028 (OC)** *Tiger* (D428) sits at Basingstoke station having just hauled both me and my son, Kevin, on the 9.10am from Waterloo on 8 April 1989. I saw this loco many times on Midland lines as unnamed D428 before it was transferred to the Western Region in 1974, gaining a name and alternating allocations between Laira (Plymouth) and Old Oak Common prior to withdrawal in 1991. Storage at Ocean Sidings (near Laira) and Exeter Riverside Yard followed; it was moved to Old Oak Common for parts removal before being cut up on site by Coopers Metals Ltd. of Cardiff in 1991.

Right: Crewe-built Class 47/4 (47844) **47556 (BR)** (47020) (D1583) sits at Basingstoke in standard B.R. Rail Blue livery awaiting the 'right away' and will soon take the line diverging to the left of the main line into Waterloo. This is because 47556 is hauling 'The Wessex Scot', bound for Glasgow Central via Reading, due to depart Basingstoke at 10.32am. The line diverging to the left goes to Reading and joins the West of England main line at Southcote Junction, just south of Reading West; it then curves around to join the other main line from Didcot, Bristol and South Wales into Reading Station. 'The Wessex Scot' was a train that ran from Poole in Dorset to Glasgow Central (and return) between 1984 and 2002. New to Cardiff Canton in 1964, 47556 also worked out of Landore (Swansea), Old Oak Common, Bescot, Laira (Plymouth) and Bristol (Bath Road), where I had previously photographed it on 20 August 1988. 47556 was reallocated to Crewe Diesel later that month and ended its days in store at Toton, being cut up by Harry Needle Railroad Company at Crewe Works in October 2004. EMUs wait for their next turn of duty in the carriage sidings. As can be seen, for the safety of staff in stations and on walking routes, the live 'third rail' is encased by kickboards. This image is dated 8 April 1989.

All aboard! Will the young man who has been admiring the locomotive, please take his seat! On 8 April 1989, Class 50 **50041 (LA) Bulwark** (D441) sits at Basingstoke station awaiting the 'Right Away'. Facing north or, more accurately, east towards Waterloo, 50041 is on an Exeter to Waterloo service and is about to haul me and my son, Kevin, to Woking. Starting life at Crewe Diesel, it moved to Bristol (Bath Road) in 1976 and later Laira (Plymouth) where it gained the above name. Previously, on 23 November 1983, 50041 was hauling the 21.35 sleeper train from Penzance when it derailed entering Paddington station at excessive speed due to driver error. 50041 slid for 100m and smashed into the ramp of platform 8, its bogies having been ripped from underneath it. After spending two months at Old Oak Common under tarpaulins, it was transported to Doncaster Works in January 1984 where it spent a year being repaired. It looks fine, now! Surviving in traffic another year until withdrawn from service in April 1990, it was cut up at Old Oak Common by Cooper's Metals Ltd. of Cardiff in 1991, along with 50005, 50009, 50024, 50028, 50032, 50034 and 50039.

CARDIFF CANTON TMD

Left: Built at B.R. Derby Works, Class 45 **45110 (TI)** *Medusa* (D73) had been unofficially named at Sheffield (Tinsley) less than two weeks prior to this image being taken on 23 August 1987. Beginning life at Crewe North in 1960, future allocations were Derby, Cricklewood, Kentish Town, various divisions of the Midland Region, Toton and Sheffield (Tinsley). I enjoy looking back through notebooks to discover where I had previously seen locos and I see that I have written down this loco as '73' at Bristol Bath Road on 1 April 1972 along with seven other 'Peaks'; I also noted it in Severn Tunnel Junction TMD on 4 August 1974; withdrawal came in July 1988. According to Greek mythology, anybody who looked into her eyes was turned to stone, so it was lucky that I was photographing her sideways on. Medusa was beheaded by Perseus on Sarpedon; 45110 was cut up by M. C. Metals Processing at Springburn Works in 1990.

Right: Old Oak Common's (47750) **47626 (OC)** *Atlas* (47082) (D1667) has put in an appearance at Canton on 23 August 1987. Built at Crewe in 1965, it went new to Landore (Swansea), later allocations being Cardiff Canton, Old Oak Common, Bristol (Bath Road), Crewe Diesel and Toton; it gained another name, Royal Mail Cheltenham in 1996. 47626 was broken up by European Metal Recycling, Kingsbury, in 2008.

Class 56 **56001 (CF)** *Whatley* was built in 1977 by Electroputere, Romania, gaining the above name in October 1987; this image is dated 23 November 1987 when 56001 had been named for almost one month. Managing to hang onto the name until withdrawal in 1996, it was scrapped on-site the following year at Cardiff Canton by M.R.J. Phillips.

Left: English Electric-built **37220 (CF)** *Westerleigh* (D6920) went new to Cardiff Canton in 1964. Further allocations were Healey Mills, Sheffield (Tinsley) and Thornaby, returning to Canton in 1974, receiving its name at the new Westerleigh Oil Terminal, Bristol in 1990; a small plate above the name was inscribed Murco. Future allocations were Immingham, Eastleigh, Stewarts Lane and Toton, where it went into store. Used for spares recovery at various sites, 37220 was eventually broken up by European Metal Recycling, Kingsbury, in 2007. This image is dated 7 March 1992.

Right: Doncaster-built Class 56 (56311) **56057 (CF)** in large logo Railfreight Grey livery had gone new to Toton in 1979, further allocations being Bristol (Bath Road), Cardiff Canton, Stewarts Lane and Motherwell where it was named British Fuels in 1996; it is still in service at the Nene Valley Railway. This image is dated 23 November 1987.

Built by English Electric in 1965, **37430 (CF)** *Cwmbran* (37265) (D6965) began life at Sheffield (Tinsley), moving on to Wath, Stratford, March, Landore (Swansea), Eastfield, Aberdeen (Ferryhill), Motherwell, coming to Cardiff Canton in 1986 when it was renumbered and gained the above name. The plate below the nameplate reads *'This locomotive was named in recognition of the support by the Cwmbran Development Corporation'*. After Canton, it was reallocated to Immingham, ending its days in store at Motherwell from where it was moved by road to European Metal Recycling, Kingsbury, in 2008 and broken up that year.

Left: Built by English Electric in 1964, **37896 (CF)** (37231) (D6931) went new to Landore (Swansea) and stayed a Landore loco for 15 years until reallocated to Bristol (Bath Road) in 1979; Cardiff Canton and then Toton were its next ports of call until going into storage at Eastleigh in 2000. Four months working in France came next followed by long-term storage in Tees Yard and Toton when returning to the UK. After more movement and storage, it was moved by road to Ron Hull of Rotherham and was cut up in 2010. This image is dated 23 November 1987.

Right: The first Class 56 built at Doncaster was (69001) **56031 (BR)** ***Merehead***, taken into stock at Toton in 1977. Future allocations were Sheffield (Tinsley), Healey Mills and Bristol (Bath Road) where it gained the above name. Cardiff Canton, Stewarts Lane, Immingham and Thornaby followed; 56031 has since been converted to 69001, the first of the Class 69s; this image of it in BR Rail Blue with yellow warning ends is dated 23 August 1987.

37429 (CF) *Eisteddfod Genedlaethol* (37300) (D6600) was the first of the last batch of nine Class 37 locos (D6600-D6608) built by English Electric in 1965. It went new to Cardiff Canton, followed by Landore (Swansea), Eastfield (Glasgow), Motherwell and back to Canton in 1986 where it was named Sir Dyfed / County of Dyfed in 1987. It kept this name for all of three months, gaining its current name at Portmadoc station later that year. Further allocations were Thornaby, Crewe Diesel, Immingham, Toton then back to Canton in 2001. Later that year it was moved to Motherwell for storage and, after six years, continued in storage at Toton; it was later moved to European Metal Recycling, Kingsbury, in 2008 and was broken up within four days of arrival. This image is dated 23 August 1987.

DONCASTER AND ENVIRONS

Built by General Motors, Canada, **66115 (TO)** heads a permanent way train next to Doncaster Royal Mail Centre. It was the last Class 66 Michael needed and I was obviously pleased to be able to send him this image taken on 9 November 1999. Going new to Toton, 66115 was later reallocated to Cardiff Canton and is still operational.

Darlington-built **08879 (DR)** *Sheffield Childrens Hospital* (D4047) went new to Sheffield (Grimesthorpe), later allocations being Sheffield (Darnall) and Sheffield (Tinsley) where it was unofficially named Earles in 1988. It gained its current name and one-off livery at Tinsley Open Day in 1996, just prior to this reallocation to Doncaster; it carries a Pudsey Bear logo on the cabside with the running number now relegated to the side of the fuel tank; a Tinsley Rose plaque is displayed centrally on the front of the loco. This image is dated 30 October 1999.

Left: Carrying InterCity Swallow livery, (91107) **91007 (BN)** was almost one year old when it failed in Doncaster station on 14 July 1989 and was removed from the front of the train by Class 08 **08885 (DR)** (D4115). 91007 would later carry the names Ian Allan, Newark on Trent and Skyfall. 08885 was new to Lostock Hall in 1962, later working out of Allerton and Hull (Botanic Gardens) where it was unofficially named Mars. Coming to Doncaster in 1985, it was withdrawn in 1993 but now works as an Industrial loco.

Right: It didn't take long for me to see all of the Class 91s and their Driving Brakes at Kings Cross but, on 14 July 1989, I still needed a couple and (91109) **91009 (BN)** in InterCity Swallow livery was a welcome 'cop'. It was also a welcome sight because, at Kings Cross, the rear end was always attached to a carriage, thus making it difficult to appreciate exactly what the rear end looked like. I took this photograph from Doncaster station; the buildings beyond the tracks form part of the Works which we always accessed via a gate in the station car park. 91009 would later carry the names Saint Nicholas, The Samaritans and Sir Bobby Robson.

On 30 March 1980, it is sad to think that Brush-built Class 31/0 **31003 (WDN)** (D5503) would soon be no more. As a boy living in Grays, I had seen this loco so many times on the ex-LT&S. Going new to Stratford in 1958, apart from an eight-month spell at March in 1959, it was never allocated to any other Depot. In all-over BR Rail Blue livery now but, back then, it was in B.R. green.

Left: It is even sadder to think that this is all that is left of another 'old friend' from the same period, Brush-built Class 31/0 **31005 (WDN)** (D5505). This image is dated 30 March 1980 but I photographed this loco leaving Tilbury Dock over 20 years previously (see above) and back then, little could I envisage witnessing the loco reduced to a pair of cabs. It had gone new to Stratford in 1958 and was never allocated to any other Depot.

Right: On 30 March 1980, **71003 (WDN)** (E5003) (E5018) has obviously been broken up recently. I had always thought that 'flame cuts' were a more recent phenomenon but, evidently not. To think that, as a boy, I made a note of seeing this loco travelling through Gravesend station in the lighter Southern Region green with a central red and white band between the cabs; where has the time gone?

The B.R.E. Works brochure displayed a shiny Deltic but, once through the gates, it was a very different story. Built in 1961 and, after a relatively short lifespan of 19 years, two nose cones are all that is left of St. Paddy on 30 March 1980, the scrapping of which had commenced on 23 February 1980. The first two Class 55 'Deltics' to be withdrawn on 5 January 1980 were 55001 (D5001) St. Paddy and 55020 (D9020) Nimbus, the latter having been scrapped one month previously; they had already been out of traffic for the best part of two years and had been heavily cannibalised for spares.

In the late 1970s, the Deltics began to be supplanted by the next generation of express trains, the High-Speed Train (HST), and the Deltics began to take on secondary roles. British Rail at this time had a general policy of not maintaining small, non-standard fleets of locomotives, and thus the class had a limited future. When the HST fleet took over on the main East Coast services it was never likely to be economic to maintain a small non-standard class of locomotive for secondary services. Thought was given to redeploying all or some of the Deltic fleet on the Midland Main Line, the Trans-Pennine route between Newcastle and Liverpool, or the North East-South West cross-country route, but all were ultimately abandoned as uneconomic, due to maintenance and crew training costs.

I photographed St. Paddy in Finsbury Park Depot when almost new and this image will be seen in a future publication, 'Railway Adventures in The Sixties and Seventies'. I also note that I saw it in Haymarket shed on 18 August 1974 and multitudinous times at Kings Cross station.

Left: English Electric-built Class 37/0 (37696) **37228 (CF)** (D6928) arrived at Doncaster Works with collision damage in 1979, presumably the end that is sheeted over, and was put into store prior to repair. Leaving ZF in 1982, it went to Sheffield (Tinsley), Thornaby, Cardiff Canton, Eastfield (Glasgow) and Motherwell. Later allocated to 'Special Projects', it went to France in 1999 and, on return in 2000, began interminable storage at Tyne Yard and various other locations. Moved by road to C. F. Booth of Rotherham, it was broken up in 2014. This image of 37228 in all-over B.R. Rail Blue is dated 30 March 1980.

Right: Landore's English Electric-built Class 37/0 (37406) **37295 (LE)** (D6995) has taken a nose-dive and ended up in Doncaster Works; this loco would later become 37406 The Saltire Society and is seen in the inset at Eastfield in 1989. New to Cardiff Canton in 1965, it also worked out of Landore (Swansea), Laira (Plymouth), Bristol (Bath Road), Eastfield (Glasgow), Motherwell and Crewe Diesel. Withdrawn in 2007, like many other diesel locos, it took a while to be put out of its misery. This job was eventually executed by C. F. Booth, Rotherham, in 2013. This image is dated 30 March 1980.

Where have the years gone? It is hard to believe that I took this photograph of a brand new **56076** outside the paint shop at Doncaster Works on 30 March 1980 (the year that John Lennon was shot dead and two years prior to the Falklands War). I obviously saw it many times in its working life but, after a lifespan of almost 30 years, it was broken up by C. F. Booth of Rotherham in 2009. I feel old.

EASTFIELD TMD

The date is 19 August 1991 and Crewe-built **08570 (ED)** (D3737) has just arrived at Eastfield from Haymarket, apparently requiring some minor repairs. New to Dalry Road in 1959, future allocations were Perth South, Thornton Junction, Dunfermline Townhill, Haymarket (Edinburgh), and Eastfield (Glasgow). Withdrawn in 1992, it was scrapped by M. C. Metal Processing in 1993. Glasgow high-rises in the background.

Built by Robert Stephenson & Hawthorns, Darlington in 1962, Class 20/0 **20122 (WDN)** (D8122) went new to Polmadie before spending a few years on the London Midland Region. Future allocations were Eastfield (Glasgow) and Motherwell followed by Sheffield (Tinsley) and then Thornaby in 1987, where it gained the name Cleveland Potash; I photographed it there as such. The nameplates were removed in 1990 when it went into long-term storage at Eastfield (Glasgow) from where it was withdrawn in July 1991. 20122 was moved to BRML Glasgow Works in 1992 and was scrapped by M. C. Metal Processing next door in 1993. This image of 20122 in Railfreight Red Stripe livery with large B.R. double arrow is dated 19 August 1991.

Built by English Electric in 1965, Class 37/4 **37410 (ED)** *Aluminium 100* (37372) (D6973) went new to Cardiff Canton, its only allocation for 20 years until it arrived at Eastfield (Glasgow) in 1985 where it gained the above name. It later worked out of Motherwell until, after much storage, it was broken up by C. F. Booth, Rotherham, in 2013. This image is dated 20 August 1989.

Above: A tale of two identities; the original Crewe-built 08731 had previously been D3899 until, whilst in Swindon Works for the fitting of train dual brakes, it was found to have a cracked frame; 08731 then exchanged identities with Crewe-built 08572 which was in the Works for scrapping. The original 08571 was renumbered 08572 and scrapped in 1983. The original 08572 then became **08731 (GM)** (08572) (D3739) in 1983, a loco that had previously been shedded at Parkhead, St. Rollox, Eastfield (Glasgow) and Motherwell. 08731 went back to Motherwell, later working out of Ayr, Inverness and Grangemouth, where it was shedded when this image was taken on 22 August 1989, despite still retaining a Motherwell shed sticker. Apparently at Eastfield for minor repairs, it was later reallocated back to Motherwell in 1993.

Right: Built by English Electric in 1967, Class 20/0 **20211 (WDN)** (D8311) was used as a parts donor (see next page) for 20206. Starting life at Thornaby, it moved to Hull (Dairycoates), Sheffield (Tinsley), Eastfield (Glasgow), Toton and Haymarket (Edinburgh). Withdrawn the previous month, it was scrapped by M. C. Metal Processing in 1993. This image of 20211 in Standard Rail Blue livery with yellow warning ends is dated 19 August 1991.

Built by English Electric in 1967, Class 20/0 **20206 (WDN)** (D8306) went new to York, later allocations being Sheffield (Tinsley), Eastfield (Glasgow) and Haymarket (Edinburgh). After receiving crash damage in February 1991, it was repaired at Eastfield using parts taken from 20211 but was still withdrawn two months later and eventually cut up. This image of 20206 in Standard Rail Blue livery with yellow warning ends is dated 19 August 1991.

English Electric-built Class 37/0 **37156 (ED)** *British Steel Hunterston* (37311) (37156) (D6856) went new to Landore (Swansea) in 1963, moving on to Polmadie, Eastfield (Glasgow) and Motherwell. Named at Motherwell station in 1986, it was then reallocated back to Eastfield in 1990; this image is dated 19 August 1991. It later worked out of Inverness, Immingham and Toton, losing the name along the way. Moved to Springs Branch (Wigan) CRDC in 1999, it was broken up in 2000.

Right: Built by the Birmingham Railway Carriage & Wagon Company, **26035 (ED)** (D5335) was new in 1959; the only Depots this loco called home were Hornsey, Edinburgh (Haymarket), Inverness and Eastfield (Glasgow). When withdrawn in 1992, preservation beckoned, and it was stored in Inverness station yard awaiting asbestos removal at M. C. Metals, Springburn Works, Glasgow. This image is dated 20 August 1989; 26035 currently resides at the Caledonian Railway, Brechin. Of equal interest is the Departmental loco in the image, ADB 968028 (27024) (D5370) which I was destined to see in a layby on the southbound A1 near Wetherby a couple of years later. I was travelling north to Redcar but, seeing it on the opposite carriageway, turned off at the next exit and drove back to photograph it (see inset). It has now been preserved and there are two very good British Rail Production videos of it working on the Caledonian Railway, Brechin, Gala Days as D5370. In B.R. green, I hasten to add; not the standard Rail Blue livery it carries here.

Above: Built by the Birmingham Railway Carriage & Wagon Company, Class 26/0 **26010 (ED)** (D5310) sits at Eastfield in standard large logo Railfreight General Grey livery and, at the risk of sounding like a broken record, it is hard for me to imagine that this loco which has had a Scottish allocation for nearly 30 years was D5310 which I saw many times in B.R. green and white livery in Kings Cross station. New to Hornsey in 1959, future allocations were Finsbury Park, Haymarket (Edinburgh), Inverness and Eastfield (Glasgow). This image is dated 20 August 1989.

Right: Built in 1964, English Electric Class 37/0 **37232 (ED)** *The Institution of Railway Signal Engineers* (D6932) was a Landore (Swansea) loco until spending most of 1978 and 1979 out of use at Doncaster Works with collision damage. After this, it worked out of Bristol (Bath Road), Cardiff Canton, Immingham and Eastfield (Glasgow) where it was given the above name at Perth station in 1990. Later allocated to Inverness and Motherwell, it endured five years of storage before being taken by road to Springs Branch (Wigan) CRDC and was cut up in 2000. This image is dated 19 August 1991.

Right: **ADB 968028** (27024) in a layby on the A1 on 3 February 1991. Incredibly interesting (to me) is that two years hence, on 13 August 1993, I would photograph 20144 being loaded onto this very same Allelys trailer (G815 GAB) in Cobra yard, Middlesbrough.

Left: Brush-built in 1966, Class 47/7 **47709 (ED)** *The Lord Provost* (47499) (D1942) spent five years working on various divisions of the Midland Region until a transfer to Bristol (Bath Road) in 1972. Later transferred to Landore (Swansea) and Laira (Plymouth), it was reallocated to Haymarket (Edinburgh) in 1979 where it gained the above name. Later allocations were Old Oak Common, Eastleigh and Crewe Diesel, prior to various sales where it gained the name Dionysos; storage at numerous sites followed, one of which, amazingly, was East Ham Electric Depot. It was cut up by Raxstar at Eastleigh Works in 2012. This image is dated 20 August 1989.

Right: Class 26/1 **26037 (ED)** (D5337) was built by the Birmingham Railway Carriage & Wagon Company and went new to Haymarket (Edinburgh) in 1959, its allocation divided between there and Inverness until it arrived at Eastfield in 1987 where it stayed (mostly in store) until withdrawal in 1993. It was cut up by M. C. Metal Processing, Springburn Works, Glasgow, in 1995. This image of it carrying Railfreight General Grey is dated 19 August 1991.

Left: Built in 1965, English Electric Class 37/**4 37407 (ED)** *Loch Long* (37305) (D6605) went new to Cardiff Canton and was one of the small batch of eight 'D66XXs' that Michael and I travelled to Wales to see when they were new, probably because we were intrigued to see Type 3s with running numbers 'D6600 to D6608' when we were so used to seeing the previous Type 3s numbered 'D67XX', 'D68XX' and 'D69XX'. This occurred because, after D6999, the next consecutive number would have been D7000 which was already the first Hymek. This loco also worked out of Bristol (Bath Road) and Landore (Swansea) before arriving at Eastfield (Glasgow) in 1984 where it gained the above name and number. This image of 37407 in InterCity livery is dated 19 August 1991. Future allocations were Inverness, Immingham and Crewe Diesel who renamed it Blackpool Tower at Euston station in 1995 and I photographed it as such in Transrail livery at Knottingley in 1997. After this it worked out of Springs Branch (Wigan) and Toton; after much storage, it was sold to Direct Rail Services in 2012 and is still operational.

Right: Class 26/1 **26028 (ED)** (D5320) was built by the Birmingham Railway Carriage & Wagon Company and went new to Leith Central for a month in 1959, after which it alternated allocations between Haymarket (Edinburgh) and Inverness. After years of storage, it arrived at Eastfield in 1987, withdrawal coming in 1991. It was cut up the following year at M. C. Metal Processing, Springburn Works, Glasgow. This image of 26028 in standard Rail Blue livery with wrap around yellow warning ends and a central running number is dated 20 August 1989.

Diesel and Electric Locomotives Around the UK in The Eighties and Nineties

Above: On 20 August 1989, with snow-ploughs covered in grime and faded paint on the bodywork, Brush-built Class **47/0 47004 (ED) (D1524)** looks a hair's breadth away from the scrap merchant. Indeed, after reallocation to Sheffield (Tinsley), it was withdrawn in 1991. However, after an inordinate amount of storage, withdrawals and reinstatements, and in a railway equivalent of the tale of the ugly duckling, 47004 became a swan; it was preserved in original two-tone green livery (standard B.R. loco green offset by a deep middle band of Sherwood Green) which, in my opinion, was the best livery ever carried by a Class 47. Its original number (D1524) was denoted at one end, 47004 at the other and it was named Old Oak Common Traction & Rolling Stock Depot in 1994. I photographed it as such at Stratford Depot on 21 August 1995 (see Stratford Depot Locomotives in The Eighties and Nineties) and at Toton Open Day on 30 August 1998 (see inset). Having gone new to Finsbury Park in 1963, I have many notes of seeing it in Kings Cross station at that time. It also worked out of Sheffield (Tinsley), Stratford, York, Gateshead, Immingham and here at Eastfield (Glasgow).

Left: **26041 (ED)** (D5341) was built by the Birmingham Railway Carriage & Wagon Company in 1959 and the only Depots this loco called home were Edinburgh (Haymarket), Inverness, Kittybrewster and Eastfield (Glasgow). Withdrawn in 1992, the loco did not arrive at M. C. Metal Processing, Springburn Works, Glasgow until 1995 and it was cut up that year. This image of 26041 carrying Railfreight General Grey livery is dated 20 August 1989.

Right: Another of the Class 37/4 selected for repainting into InterCity livery is English Electric **37431 (ED)** *Bullidae* (37272) (D6972). New to Cardiff Canton in 1965, it also worked out of Laira (Plymouth), Eastfield (Glasgow) and Motherwell. Back at Canton in 1986 it was renumbered and named Sir Powys / County of Powys, keeping it until 1991 when it gained its current name whilst allocated to Immingham. It had been an Eastfield charge for three months when this image was taken on 19 August 1991. It then worked out of Inverness and Motherwell before storage at various sites, its final journey being to Springs Branch (Wigan) where it was cut up in 1999.

A loco also seen in the Inverness section of this book, on 19 August 1991, (47776) **47578 (WDN)** (47181) (D1776) is now withdrawn and in a scrap line alongside 20122 (ex 'Cleveland Potash'). No doubt their nameplates are on their way to an auction site, just rusty oblong marks left on the locos. However, any lumps in throats are premature because, a year later, 47578 was reinstated; it was reallocated to Crewe Diesel, given B.R. Parcels livery, a new number and named Respected; it can be seen in this guise in the Euston station section of this book. Withdrawn again in 2004, it is currently stored, non-operational, for West Coast Railways. As you may have gathered by now, it has always fascinated me how locos could change their identities multiple times. The inset shows this loco in a previous life passing Stratford Depot in 1969 as D1776.

Brush-built in 1966, Class 47/7 **47702 (ED) *Saint Cuthbert*** (47504) (D1947) spent the first six years of its life on the London Midland Region before being transferred to Landore (Swansea) in December 1972. It became a Haymarket (Edinburgh) loco in 1978 and received the above name. After this, it spent time at Eastfield (Glasgow), Old Oak Common, Eastleigh and Crewe Diesel. This image is dated 22 August 1989 and 47702 carries ScotRail livery and I was to photograph it again six years later all but a day on 21 August 1995 at Stratford TMD; by that time 47702 had changed its livery to Railfreight Triple-Grey, its name to County of Suffolk and was shedded at Stratford (see Stratford Depot Locomotives in The Eighties and Nineties).

STRATFORD STATION

Left: On 9 September 1999, (69005) **56007** in Transrail livery heads a train of 'cargowaggons' through Stratford station. Surely an appropriately numbered loco for the 70th anniversary (1953-2023) of the year Ian Fleming introduced James Bond to the world. I remember reading the Pan paperbacks before any film was thought of, never foreseeing the marketing phenomenon this fictional character would become. 56007 survived catching fire in 1994, later worked in France and has recently been converted to a Class 69, number 69005.

Right: Extremely interestingly, this is the loco that makes up most of the 'new' D5910 'Baby' Deltic project and the unique Napier T9-29 'Baby' engine was lowered into 37372 in 2009; the loco was later sliced in two with a view to shortening it. I say 'most' because one nose cone was also used from 37891 (pictured on page 30 of Stratford Depot Locomotives in The Eighties and Nineties). But on 7 July 1999, Robert Stephenson & Hawthorns-built **37372 (TO)** (37159) (D6859) in Mainline Freight Aircraft Blue livery ploughs a lone furrow along Stratford station. Beginning life at Cardiff Canton in 1963, later allocations were Landore (Swansea), Polmadie, Haymarket (Edinburgh), Eastfield (Glasgow), Bristol (Bath Road), Stratford, Toton, Eastleigh, Old Oak Common and Crewe Diesel. I wonder if the 'Baby' Deltic project noticed that it had two odd buffers? I also trust that readers appreciate the way I clear the platform of people before taking a photograph!

47339 (FD) (D1820, **47157 (FD)** *Johnson Stevens Agencies* (D1750), **86611 (CE)** *Airey Neave* (86411) (86611) (86411) (86311) (86011) (E3171), **86610 (CE)** (86410) (86010) (E3104) travel through Stratford station on 21 July 1999; the train locos were the two Class 86s, the two Class 47/1s being moved from one storage location to Crewe Diesel for continued storage. And three of the four are doomed! Class 47/3 47339 was withdrawn two months after this image was taken and, after years of storage, was eventually cut up by C. F. Booth of Rotherham in 2005. Class 47/0 47157 had already survived colliding with 09004 at Paddock Wood in 1993 but was eventually cut up by C. F. Booth of Rotherham, in 2004. 86611 was destined to be badly damaged in a collision at Shallowford in 2003 when, whilst in tandem with 86631, the pair ran into the rear of a stationary Isle of Grain-Trafford Park Freightliner, 86611 ending up in the rear cab of 86631. 86611 was cut up by the Harry Needle Railroad Company at Crewe Works two years later. 86610 managed to stay out of trouble and is still in one piece (just); it was withdrawn in 2020 and is currently stored at Crewe Basford Hall.

BARRY STATION AND TMD

With the old Barry Northlight steam shed in the background, English Electric-built Class 37/7 **37799 (CF)** (37061) (D6761) in large logo Railfreight Grey livery, heads a train of empty hoppers on 23 August 1987. New to Thornaby in 1962, it later worked out of Gateshead and Cardiff Canton where it was named Sir Dyfed / County of Dyfed in 1986; it moved to Motherwell in 1996 before Toton and Crewe Diesel beckoned. With nameplates removed, it left for Spain in 2001 via the Channel Tunnel and was broken up at Vilafranca Del Penedes in 2011.

Barry engine shed (88C) was located on the south side of Barry Town station; it closed to steam in 1964 but at this time was still used as a wagon repair depot and diesel stabling point. One of the last Class 08s constructed, Darlington-built Class 08 **08942 (CF)** (D4172) in all-over B.R. Rail Blue with red buffer beams and yellow and black 'wasp' ends went new to Cardiff Canton in 1962; it then moved to Newport (Ebbw Junction), Landore (Swansea), Bristol (Bath Road) and, in 1986, ending up where it began, at Cardiff Canton. This image is dated 23 August 1987. The yard is a feast for modellers, from the row of buffers to old blue oil drums; the green footbridge in the background (looking suspiciously like the one modelled by Hornby) shows how close the shed was to the station.

GATWICK AIRPORT STATION

Built by English Electric at Vulcan Foundry, Class 73/2 **73210 (SL)** *Selhurst* (73116) (E6022), seen here in the later InterCity Gatwick Express livery, went new to Stewarts Lane in 1966 and stayed there, apart from a year at Eastleigh in 1967. Withdrawn in 2002 and out of use in 2003 in a bad state of repair, it has since been preserved.

Built by English Electric in 1966, Class 73/1 **73114 (SL)** *Stewarts Lane Traction Maintenance Depot* (E6020) carries Mainline livery and is likely to be on a Brighton-Manchester/Glasgow train, standing in for a failed Class 47, which I am informed often happened. 73114 was first allocated to Stewarts Lane and later, Hither Green and Eastleigh. After derailment damage, it went into long-term storage at Old Oak Common and Peak Rail, Derbyshire. Withdrawn in 1999, it is now preserved.

Left: English Electric-built Class 73/1 (73235) **73135 (SL)** (E6042) went new to Stewarts Lane in 1966. On 10 April 1991, it was renumbered 73235 for use on Gatwick Express duties but, as can be seen, it was already performing them a couple of months prior to this on 17 February 1991 carrying INTERCITY Executive Swallow livery. It displays its new number in the other image on this page dated 5 August 1994.

Right: 73135 has now been renumbered and carries InterCity Gatwick Express livery. Whilst hauling the 05.20 Gatwick to Victoria Gatwick Express in June 2003, Class 73/2 **73235 (SL)** (73135) (E6042) and coaches became derailed between Redhill and Sand Tunnel on the Quarry line due to defective track. Rerailed later in the day, 73235 was moved to Stewarts Lane Depot for assessment. Due to the damage received, the cost of repair was deemed too expensive and the loco was withdrawn. However, overhauled in 2005, it was leased to South West Trains where it ended its days with them shunting at Bournemouth Depot; it is now preserved.

English Electric-built Class 37/0 **37106 (SF)** (D6806) approaches Gatwick Airport station with a permanent way train. It went new to Sheffield (Tinsley) in 1963, future allocations being Gateshead, Healey Mills, Immingham, Motherwell, Eastfield (Glasgow), Inverness, Stratford and Eastleigh. When photographed here on 7 July 1995 in 'Dutch' livery, 37106 was in the middle of a four-year stint at Stratford. It was scrapped at Springs Branch (Wigan) CRDC in 2000.

Left: English Electric-built Class 73/2 **73201 (SL) *Broadlands*** (73142) (E6049) went new in 1967 to Stewarts Lane. Gaining the above name in 1980, it was renumbered in 1988 for use on Gatwick Express duties and retained the nameplates. It remains in service.

Right: With Gatwick Express duties now a distant memory, 73201 lost the livery but at least here at Hoo Junction in 2015 it now has a yellow roof to be proud of.

Built by English Electric in 1966, Class 73/2 (73952) **73211 (SL)** (73113) (E6019) enters Gatwick Airport station on 5 August 1994 carrying the later Gatwick Express livery. It was always a Stewarts Lane loco except for a year at Eastleigh in 1967 and until 1991, it carried the name County of West Sussex. After withdrawal, this loco was heavily stripped of parts at Stewarts Lane until being purchased by RVEL (Railway Vehicle Engineering Ltd.) It was rebuilt to become the 'Ultra73' prototype locomotive, was unveiled as 73952 in 2014 and was named Janice King in 2016.

Built by English Electric in 1966, Class 73/2 (73963) **73206 (73A)** *Gatwick Express* (73123) (E6030) and Class 73/2 (73965) **73208 (SL)** *Croydon 1883-1983* (73121) (E6028) [opposite page] are photographed on 7 July 1995. I was destined to capture their images again at Eastleigh on 24 June 2007 when, as can be seen to the right, 73206 had turned into Lisa and 73208 had become Kirsten. That is not the end of the story, though; they have been renumbered for a fourth time as 73963 and 73965 respectively. Improvements by RVEL enable a Class 73/9 locomotive to provide the same power operating under diesel as it does via the third rail, whilst being cleaner, quieter and more fuel efficient.

IMMINGHAM TMD

Left: Although now a Sheffield (Tinsley) loco, Class 47/3 **47379 (TI)** *Total Energy* (D1898) began its life here at Immingham in 1965 and came back for another 11 years in 1973. These were the only two sheds it was allocated to apart from very brief spells at Cardiff Canton and Bescot. Built by Brush in 1965, it was cut up at Springs Branch (Wigan) CRDC in 1999. It sits at Immingham between duties on 1 July 1989. Before the Humber Bridge was built, Michael and I went to the Hull sheds first (Dairycoates and Botanic Gardens) and then made the journey across the Humber to New Holland by ferry, driving down the platform at New Holland Pier station. I note that we did that on 2 June 1973 and 2 September 1979 where I note that the ferry from Hull was B.R. ferry 'Farringford', arriving at New Holland at 6pm. The drive to Immingham was 21 miles and I note that we arrived at 6.55pm and proceeded to get around the shed without any problems; there were 34 locos on shed, five of which were 'cops'. However, moving on to 1 July 1989 when these photographs were taken, the ferry had long-since been defunct, having stopped on 24 June 1981, the day the new Humber Bridge opened.

Right: Built at Vulcan Foundry, Class 37/0 **37275 (IM)** *Stainless Pioneer* (D6975) went new to Cardiff Canton in 1965 and remained there until reallocated to Immingham in 1988. It went back to Canton for a brief spell four years later before moving on to Inverness (where it gained the name Oor Wullie), Motherwell, Bescot and Toton; it is now preserved and operational at The Dartmouth Steam Railway. This image is dated 1 July 1989. Michael and I did not visit Immingham many times because it was not the easiest shed to get to, out on a limb near the south-west corner of Immingham Docks and not close to any other sheds. In addition to that, it was not an easy shed to get around once you were there so there was a real chance that the effort to get there might be in vain. The whole area was always a mysterious place to visit, almost like something out of Quatermass; it seemed permanently dark, even at midday, and there was always a sort of sulphurous smell in the air.

Brush Class 31/1 **31273 (IM)** (D5803) was involved in two accidents during service. In 1977, it overran the end of the goods loop at Hanwell and crashed off the end of the line and onto its side; it remained halfway down the embankment for nearly a month, after which it was repaired before being returned to service. The second occasion was in 1997; it was coupled with 31275 and 31410 and being hauled by 31201 when the quartet was derailed near Carnforth which, forgive the pun, was the end of the line for 31273. It was stored at various locations until withdrawal in 2001, after which it was scrapped by Booth Roe Metals, Rotherham. Beginning life at March, other locations were Lincoln, Immingham, Old Oak Common, Bristol (Bath Road), Crewe Diesel and Bescot. This image is dated 1 July 1989. On that date, I noted down 31 numbers but I also noted 'not all numbers taken'. This invariably means that we were 'thrown out' by the foreman and, on this occasion, I remember we were. However, my time wasn't totally wasted on that visit as I note I copped Immingham's 08508.

SHENFIELD STATION

The view is different today; a major track reconfiguration has taken place in order to accommodate sidings for Crossrail trains. As a boy, my first sighting of Class 31/1 **31187 (SF)** (D5610) was in Finsbury Park Diesel Depot on 15 May 1965, resplendent in B.R. green and numbered D5610; almost 25 years later on 18 January 1989, it is unrecognisable as its former self as it makes its way light-engine through Shenfield station in large logo Railfreight Red Stripe livery. Of course, almost another 35 years have passed since this photograph was taken so if anybody out there has discovered where time goes, please let me know; I want some of it back, price no object. 31187 went new to Finsbury Park in 1960, further allocations being Gateshead, March, Immingham, Stratford and Toton, withdrawal coming in 1995. After much storage, it was finally broken up by T. J. Thomson of Stockton.

Left: In the deep midwinter, Cardiff Canton Class 37/0 (37891) **37166 (CF)** (D6866) travels light engine through Shenfield station on 15 January 1987; within a few months it would become a more familiar sight on the line when reallocated to Stratford. Coincidentally, Cardiff Canton was the first allocation for this locomotive nearly 25 years previously when it left the Works of Robert Stephenson & Hawthorns in 1963; after that, it worked out of Healey Mills, Thornaby and Immingham and came to Stratford for the first time in 1970. Allocated to 'Special Projects' in 1999, it moved to France but, on return, in 2000, went into long-term storage until it was finally put out of its misery when broken up by Ron Hull Jnr. of Rotherham in 2010. This loco also appears in Stratford Depot Locomotives in The Eighties and Nineties when, a year or so later, it had been reclassified, renumbered 37891 and given Railfreight Petroleum livery.

Right: English Electric-built Class 37/0 **37032 (TI)** (37353) (37032) (D6732) had gone new to Hull (Dairycoates) in 1962. Further allocations were Healey Mills, Sheffield (Tinsley), Thornaby, Gateshead, Immingham and it gained the unofficial name of Mirage in 1992. Withdrawn in 1994, it is now preserved at the North Norfolk Railway. On 12 July 1990, 37032 is seen carrying Railfreight Red Stripe livery in the siding at Shenfield station after developing a fault (the same siding that has been home to 'The Shenfield Shark' for a great number of years).

Left: A pair of Class 90/1 mixed-traffic locomotives **90144 (FE)** and **90150 (FE)** double-head through Shenfield with a freightliner train on 20 July 1999. The Class 90 is a modernised derivative of the Class 87 loco. Prior to the ex-G.E. line being allocated its own batch of Class 90s (90001 to 90015) for the Liverpool Street to Norwich passenger expresses, any '90' could have hauled you. However, I still sometimes got lucky when necessity demanded that a '90' was required on loan from another company such as Freightliner or EWS and I have notes of being hauled by 90019 Penny Black, 90034, 90036 and 90039. Sometimes, these visitors would stay for a few days, sometimes longer. For instance, I was hauled by 90036 twelve times.

Right: Class 86/6 **86639 (CE)** (86439) (86039) (E3153) built by English Electric at Vulcan Foundry and **86615 (CE)** *Rotary International* (86415) (86615) (86415) (86315) (86015) (E3123) built at British Rail Doncaster Works, make their way through a busy Shenfield station with a Coatbridge to Felixstowe North Intermodal. Both locos had been allocated to ACL and Willesden, prior to Crewe Electric. Withdrawn in 2005, 86615 succumbed to the cutter's torch courtesy of Ron Hull, Rotherham, in 2007 whereas 86639 is still operational. This image is dated 5 August 1999 and, when I photographed them, I had no idea that on 2 September 1967, 86639 (as E3153) had pulled Michael and me from Euston to Liverpool Lime Street.

Both built by English Electric at Vulcan Foundry, **86602 (CE)** (86402) (86002) (E3170) and **86607 (CE)** *The Institution of Electrical Engineers* (86407) (86007) (E3176), approach Shenfield station with a Felixstowe North to Trafford Park freightliner service; pairs of the Class 86 were used until the advent of the Class 70, and a long time after that. Both locos had been allocated to ACL and Willesden, prior to Crewe Electric. 86602 was withdrawn in 2004 and broken up by Sandbach Car & Commercial Dismantlers Ltd. whereas 86607 is still operational. This image is dated 8 July 1999.

Left: Built at Doncaster in 1965, **86232 (WN)** *Harold MacMillan* (E3113) went new to 5H, further allocations being ACL, Willesden, Ilford and Norwich (Crown Point). When Class 86 locomotives were introduced onto the ex-G.E. line, I determined to be hauled by as many of them as I could, even if only for short journeys; on this date, Saturday 20 February 1988, I had driven from Galleywood to Shenfield station (where I could park on the road until 10am) and 86213 Lancashire Witch hauled me on the 7.45am (ex-Liverpool Street) to Chelmsford. This new resolution fortunately coincided with my discovery of **'The European'** which meant that I simply changed platforms at Chelmsford and was hauled back to Shenfield at 08.32 by whatever Class 86 was on **'The European'**; it was usually a loco that had not yet been transferred or loaned to the ex-G.E. line and it was exciting, standing on platform one at Chelmsford station, wondering what the loco would be. **'The European'** was a train that ran from Edinburgh Waverley and Glasgow Central to Harwich Parkeston Quay and return between 1983 and 1988. This was the first time I had seen 86232 on 'my line', although later it would be reallocated to Ilford. The time of the train meant that I was back at Shenfield and photographing the loco (I always positioned myself towards the front of the train before arrival at Shenfield) long before any traffic wardens took an interest in my car (however, had I encountered one, I would simply have pointed out what it said on the sign – 'fine for parking'). These were good days, sometimes being pulled by two locomotives that had never previously hauled me, in the space of an hour.

Right: The date is Friday 26 February 1988 and 1965 English Electric-built **86225 (WN)** *Hardwicke* (E3164) sits at Shenfield heading **'The European'**. On this occasion, I had been hauled on the 07.45 Shenfield to Chelmsford by 86249 County of Merseyside. At Chelmsford, I eagerly awaiting the arrival of **'The European'** and I was not disappointed because, although I had previously photographed 86225 at Euston, this was the first time I had seen it on 'my line', as I called it, and I cannot find a note of seeing it again after this date. The guard gives the 'Right Away' and the driver acknowledges (albeit feebly) with a hand from his cab window. I suspect that some readers will be wondering if I achieved the goal I set myself in the previous narrative regarding being hauled by every Class 86; sadly, the answer is no; all but two.

Left: The next day, Saturday 27 February 1988, I was hauled on the 07.45 Shenfield to Chelmsford by 86237 Sir Charles Hallé, a loco in whose cab I had travelled from Liverpool Street to Chelmsford a few weeks previously; but more of that in another book. The loco on **'The European'** was 1965 Doncaster-built **86231 (WN)** *Starlight Express* (E3126) which went new to 5H, further allocations being ACL, Willesden and Longsight (Manchester). I had never seen it on 'my line' before and I never did again. Withdrawal came in 2002 but, after long-term storage, it was exported to Bulgaria in 2016 and, confusingly, renumbered 85005.

Right: On 2 April 1988 I went through my now familiar ritual of driving to Shenfield and purchasing a day-return to Chelmsford; I caught the 07.45 as usual and 86215 Joseph Chamberlain arrived at Chelmsford in good time for me to transfer from platform 2 to platform 1 and anticipate what loco would be hauling the 08.32. When **'The European'** arrived, it was headed by (86702) **86260 (WN)** *Driver Wallace Oakes G. C.* (86048) (E3144); built at Vulcan Foundry in 1966, it went new to ACL, further allocations being Willesden, Longsight (Manchester) and Norwich (Crown Point). In 2009 it was renumbered 86702 and renamed Cassiopeia; it never worked much and was put into store at Ilford Depot who later used it as a static carriage heater. Withdrawn in 2015, it was exported to Bulgaria in 2016 and renumbered 85002. On arrival at Shenfield, I ran to the front of the train to obtain a photograph; the driver is giving me what my maternal grandmother would have called 'an old-fashioned look'. Interestingly, as I still had to go work that day for an early shift at the casino, I was hauled by Joseph Chamberlain twice, the second time being the 21.30 from Liverpool Street to Chelmsford.

A 'Scotty Dog' at Shenfield. Eastfield's Class 47/4 (57304) (47807) **47652 (ED)** (47055) (D1639) sits at Shenfield on **'The European'** on Saturday 9 April 1988 awaiting the 'Right Away' from the guard. On this occasion, I had boarded the train at Chelmsford, jumping off at Shenfield to photograph the locomotive. With a departure time of 08.32 from Chelmsford, I have a note that on this day it was running late, perhaps because of the snow. I have also noted that this was the first time I had seen **'The European'** hauled by a diesel as opposed to a Willesden Class 86 electric. Crewe-built in 1964, 47652 went new to Cardiff Canton, further allocations being Landore (Swansea), Old Oak Common, Bescot, Stratford, Gateshead, Eastfield (Glasgow), Bristol (Bath Road) and Crewe Diesel. It was later rather disappointing to discover that 47652 had been shedded at Stratford in another life and so must have travelled though Shenfield as 47055 on many occasions; a shame, this Scottish loco seemed so exotic at the time, heading **'The European'** with a big 'West Highland Terrier' logo on the side. Knowing that the days of **'The European'** were numbered, on Saturday 9 April 1988, I boarded the train at Chelmsford, jumped off at Shenfield for this photograph and quickly hopped back on for the journey to Camden Road.

CAMDEN ROAD STATION

Obviously, the date is still 9 April 1988 and **86204 (WN)** *City of Carlisle* (E3173) was about to pull me from Camden Road to Watford Junction on a snowy morning. It was built by English Electric at Vulcan Foundry in 1965, allocations being ACL, Willesden (where it gained the above name in 1978), Norwich Crown Point and Longsight Electric Depot (Manchester). Stored at Springburn Works in 1978, it lay derelict until moved to Immingham Railfreight Terminals in 2002 and was broken up on site the following year by staff from Easco. 86204 had previously pulled me three times in 1987 on the 21.30 Liverpool Street to Chelmsford (25 June, 9 July and 18 September).

LIVERPOOL STREET STATION

16 April 1988 was 'Network Day' which allowed travel anywhere on Network SouthEast for £1. Obviously, this was an opportunity too good to miss so, together with my son, Kevin, and his friend, I drove to Shenfield (no parking restrictions then on a Saturday) and we left Shenfield station on **'The European'** behind 47436 (CD) (D1552). Due to a points failure at Stratford, 47436 took us into Liverpool Street instead of turning off at Stratford for the North London line. This was in order for the train to access the third part of the triangle, Carpenter's Road curve, that gave access to the North London line. This was a bonus for us as **47547 (CD)** (D1642) coupled onto the other end and hauled us out of Liverpool Street to Camden Road via Stratford's Carpenter's Road curve. These two images show 47537 at Liverpool Street and on Carpenter's Road curve, the latter obviously taken from the carriage window, something impossible to do, these days. I wonder if I was lucky enough to be aboard **'The European'** the only time it visited Liverpool Street on its way to Scotland; and what must have other passengers made of it? The passengers who began their journey at the front of the train were now at the back. I noted that D200/40122 (it carried both numbers) was 'back home', sitting outside Liverpool Street station with a headboard stating 'The Whistler Farewell'. This 'special' was a trip to Norwich behind a 'whistler' (Class 40) celebrating 30 years since they took over from 'Britannias' on the Liverpool Street to Norwich service in 1958.

CARPENTER'S ROAD CURVE, STRATFORD

CAMDEN ROAD STATION

Obviously still 16 April 1988, after changing locos for the third time in an hour, **85014 (CE)** (E3069), seen here at Camden Road, pulled us to Watford Junction. I was surprised at this choice of motive power because I had previously been told by a railwayman at Stratford that the Class 85 was not allowed on the North London line. A year after this image was taken, it suffered a severe fire in Euston station and was withdrawn in 1989. After storage at Crewe Electric, in 1992, 85014 was hauled to M. C. Metal Processing, Springburn Works, Glasgow, and scrapped. Built at Doncaster in 1961, it had gone new to ACL, future allocations being 5H and Crewe Electric. The live third rail that can be seen was for the EMUs on the Broad Street to Richmond service. Earlier, there would also have been a fourth live rail for the older London Midland EMUs on that service and the Euston to Watford service; Michael and I used to travel on the latter from Euston to Willesden Junction. The reflection of the Mark 2 carriage door on 85014's bodywork is the closest that any Class 85 ever got to carrying Intercity livery; none ever did.

WATFORD JUNCTION STATION

Left: Arriving at Watford Junction, we saw Bescot's Brush-built Class 31/4 **31412 (BS)** (31512) (31412) (D5692) in standard all-over B.R. Rail Blue with yellow warning ends, trapped in a platform by a red flag and a yellow lamp; the date is still 16 April 1988. 31412 went new to Sheffield (Darnall) in 1961. Future allocations were March, Sheffield (Tinsley), Ipswich, Norwich (Thorpe), Immingham, Bristol (Bath Road), Old Oak Common, Bescot, Crewe Diesel, Springs Branch (Wigan). It was sold to a private buyer in 2001 with the intention of operating under the Fragonset Banner after refurbishment. 31412 was stored at Barrow Hill roundhouse in 2002 until withdrawn in 2006 when it was moved to Ron Hull Junior, Rotherham, and broken up that year. I assume we travelled back to Euston on a local train because I have no note of being loco-hauled and I remember we saw Duncan Goodhew both on the platform at Euston and on the Northern Line underground platform. After visiting The Trocadero, Virgin Megastore and Hamleys, it was back to Euston to be hauled to Watford Junction by 86102 (WN) *Robert A. Riddles.*

Right: Doncaster-built **86419 (WN)** (86319) (86019) (E3120) then took us to Mitre Bridge and Kevin (on right) and his friend are hanging out of the window wondering if I will have time to take a photograph and make it back before the train departs; the driver appears fascinated with the traffic cone. New in 1965 to 5H, later allocations for 86419 were ACL, Willesden and Crewe Electric; in 1990 it was named Post Haste – 150 years of Travelling Post Offices at Stratford DRS. It was broken up on site at Crewe Electric Depot by Sandbach Car & Commercial Dismantlers Ltd. in 2003. 47508 (OC) S.S. Great Britain took us to Liverpool Street from Mitre Bridge; I assume there was still points trouble at Stratford. The date is still 16 April 1988. 47436 then took us to Shenfield, the seventh loco to haul us that day.

SHENFIELD STATION

Left: On Friday 6 May 1988, I travelled behind 86229 Sir John Betjeman to Chelmsford from Shenfield on the 07.45. After changing platforms, I went through the usual pangs of anticipation, wondering what would be hauling **'The European'**. As it approached, my first impression was that the loco could do with a repaint but all was forgiven when I saw it was 1965 English Electric-built **86217 (WN) *Halley's Comet*** (86504) (86217) (E3177), a Class 86 I had not seen on the ex-G.E. before and one I had been waiting to be hauled by. New to 5H, future allocations were ACL and Willesden; it had previously been named Comet and later became City University. 86217 was exported to Hungary in 2013 from Immingham Docks. All of the class were allocated new to 5H (the code used for all WCML electrics from October 1963 to January 1966).

Right: The next day, Saturday 7 May 1988, Doncaster-built (86610) **86410 (WN)** (86010) (E3104) sits at Shenfield station prior to pulling me and my son Kevin on the 08.46 to Watford Junction; This was the first Class 86/4 I had seen hauling a passenger train on the Liverpool Street line (as I called the ex-G.E.) and, therefore, it was the first Class 86/4 I had travelled behind on that line. I had taken Kevin with me because **'The European'** had only one more week to run, ending an interesting, albeit short-lived, part of railway history; I'm not sure though if, over 35 years later, it is one of his abiding memories. The Class 86/4 was a rarity on the line at this time but, within a couple of years, it would be renumbered 86610 and would doubtless double-head with another of the class through Shenfield many times on Freightliners. New to 5H in 1965, further allocations were ACL, Willesden, Crewe Electric and Crewe Freightliner. At the time of writing, this loco is in storage at Crewe Basford Hall.

MITRE BRIDGE JUNCTION

This is where my memory refuses to synchronise with what I wrote at the time. We were pulled back to Mitre Bridge from Watford Junction by 87031 Hal o' the Wynd and I noted that we 'jumped train at Mitre Bridge while the locos were being changed'. This seems a rather foolhardy thing to do as I was accompanied by my son, and I have no memory of actually doing it. I cannot remember if there was a small platform at Mitre Bridge but, it cannot be denied, 47602 and 47622 appear to have been photographed from ground level. I also noted that I had already been pulled by 47602 and that, had it been 47622 taking our train forward, I would have stayed aboard. In addition, I also noted that we later saw 47602 in Willesden Electric Depot so, I can only assume that we had walked to the Electric Depot from Mitre Bridge.

Left: Obviously, it is still the 7 May 1988 and Brush-built Class 47/4 (47782) (47824) **47602 (OC)** *Glorious Devon* (47185) (D1780) sits at Mitre Bridge waiting to couple onto our train when 87031 came off. Going new to Sheffield (Tinsley) in 1964, future allocations were Stratford, Leeds (Holbeck), Bristol (Bath Road), Landore (Swansea) and Cardiff Canton, gaining the above name in 1985. It then went to Old Oak Common, Crewe Diesel before being stored at various locations. In 1987, it was moved by road from Old Oak Common to T. J. Thomson of Stockton and broken up. This loco had previously hauled me from Shenfield to Liverpool Street (as D1780, then unnamed) on 16 March 1973.

Right: Brush-built Class 47/4 (47841) **47622 (OC)** *The Institution of Mechanical Engineers* (47134) (D1726) also waits for a train on the same date. New to Cardiff Canton in 1964, it also worked out of Bristol (Bath Road), Landore (Swansea), Old Oak Common, Eastfield (Glasgow), Crewe Diesel and Toton. It had obviously already lost this name when christened Spirit of Chester in 1999. It was withdrawn from service in 2014 and is currently stored. I had previously photographed this loco on 4 September 1987 in Liverpool Street station.

WILLESDEN ELECTRIC DEPOT

Left: Obviously, it is still 7 May 1988 and about 15 minutes previously, Class 87/0 **87031 (WN)** *Hal o' the Wynd* had hauled us from Watford Junction to Mitre Bridge and now it waits to manoeuvre onto Willesden Electric Depot. The footbridge above the loco has been used in a few television dramas, one being an episode of New Tricks called 'Good Morning Lemmings' which aired in October 2010. Three detectives played by Dennis Waterman, Alun Armstrong and Amanda Redman, stand by the footbridge before climbing the steps; they then stand on the footbridge that Michael and I have used many times on our way to Old Oak Common. All three look down at the tracks to where a 'dead body' had been found and then walk across another section of the bridge where Willesden Junction High Level station can be seen in the background. 87031 was built at Crewe in 1974, going new to Willesden, its only allocation. The 'dead body' in New Tricks actually drew breath for two months longer than 87031 which, unfortunately, became a dead body (and chassis) after withdrawal in 2005 when it was moved by road to European Recycling, Kingsbury, and cut up in August 2010.

Right: Built in 1973, Class 87 **87007 (WN)** *City of Manchester* went new to Willesden and was not named from new; however, at Manchester Piccadilly station in 1977, it took the name from the one previously bestowed upon LMS Princess Coronation Class 46246. Had 'TOPS' not coincided with the year the loco was built, 87007 was to have been numbered E3207. In 2006 it was rumoured that 87007 was to be named Roger Moore but, for some reason, this idea was abandoned. I wonder if it was because 87007 was unable to raise one of its windscreen wipers in the way Roger Moore could raise an eyebrow? I can understand the reasoning behind naming 87**007** after a James Bond actor but 87008 (also on shed) was to have been named Timothy Dalton which does not make the same sense; maybe that idea was also abandoned because somebody suddenly realised that he hadn't been 008. However, a train full of passengers were definitely both shaken and stirred when, on 16 February 1980, a broken welded rail at Bushey caused a train hauled by 87007 to derail at 96mph, injuring 19 passengers. 87007 was exported to Bulgaria in 2008. This image is dated 7 May 1988.

Willesden Electric Depot to Old Oak Common was no more than a ten-minute walk, a journey similar to the one I had made countless times from Willesden steam shed, and photographs taken there are included in the Old Oak Common section of this book.

SHENFIELD STATION

"If you miss this one, there'll never be another one", is a line sung by Johnny Duncan in the song, Last Train to San Fernando. Fast-forward one week and it is Saturday 14 May 1988, the last day of **'The European'** service; I know it is a special occasion, but did it warrant George W. Bush occupying the driver's seat? Surely it is either his Spitting Image puppet or a very lifelike doppelganger, but extremely heart-warming that he has made his own headboard to mark the end of the life of **'The European'**. The last doors are being slammed, no doubt the guard has shouted, "All aboard!" and Doncaster-built (86604) 86404 **(WN)** (86004) (E3103) is about to depart Shenfield station on the 08.46 to Edinburgh and Glasgow (and I'm about to hop on board and travel as far as Watford Junction). I had photographed 86404 travelling in the opposite direction the previous evening at Margaretting en route to Harwich, as can be seen later in this book. New to 5H in 1965, further allocations were ACL, Willesden, Crewe Electric and Crewe Freightliner. At the time of writing, this loco is in storage at Crewe Basford Hall. I would have originally 'copped' this loco in Willesden Electric Depot or Euston station when it carried the original livery of Electric Blue with raised silver numbers. I was destined to see this loco again at the 1995 Basford Hall Rail Fair (an 'open day' to you and me) as 86604 carrying Railfreight Distribution livery, a photograph of which can be seen later in this book.

WATFORD JUNCTION STATION

After alighting at Watford Junction, 86404 continued on its way and we changed our allegiance to Crewe-built Class 87 **87028 (WN)** *Lord President* and travelled behind it back to Mitre Bridge. 87028 had gone new to Willesden in 1974 and was named four years later. It was taken out of use in 2003, nameplates removed, and stored. However, due to loco shortages, it was put back into service for a month without nameplates. Sold to the Bulgarian Railway Company, it left Hull Docks for Rotterdam in 2008 and was then taken by barge up the Rhine to Bulgaria. It is still operational. Obviously, it is still 14 May 1988.

MITRE BRIDGE JUNCTION

About half an hour later, we arrived at Mitre Bridge and 87028 came off to be replaced by Brush-built Class 47/4 **47540 (OC)** (47975) (47540) (D1723). In the first photograph, I am leaning out of the window of the first carriage to obtain a shot of the situation while 87028 is uncoupled; 47540 waits to back on to our train. In the second image, 87028 has uncoupled and moved off; the points have changed and 47540 still waits patiently for 87028 to come back and cross over onto the other line towards Willesden. Mitre Bridge Junction signal box is visible just beyond the level crossing. It was of London & North Western Railway design and was located in the divergence of the West London line and the Mitre Bridge curve to Willesden Junction High Level; it closed on 14 October 1990. The unusual semaphore signals are 'shunt signals' or 'calling-on' signals, and, in the first picture, 87028 has been advised it may proceed with caution so far as the line is clear, or to the next signal. In the second image, the signal has returned to 'danger'.

87028 has now set back and is on its way to Willesden Electric Depot. Any minute, 47540 will move out of the siding, reverse, and couple onto our train ready to haul us to Kensington Olympia station; obviously, it is still 14 May 1988.

MITRE BRIDGE JUNCTION

KENSINGTON OLYMPIA STATION

Left: We alighted at Kensington Olympia after being hauled by 1964 Brush-built Class 47/4 **47540 (OC)** (47975) (47540) (D1723) on the 10.15am from Mitre Bridge on 14 May 1988. Beginning its working life at Bristol Bath Road, 47540 also worked out of Cardiff Canton, Sheffield (Tinsley), Crewe Diesel, Landore (Swansea), Old Oak Common and, after renumbering to 47975 in 1990, it was named The Institution of Civil Engineers the following year; it was renumbered back to 47540 in 1995 before going into long-term storage. Withdrawn in 1998, it was moved by road to T. J. Thomson, Stockton, in 2016. I had previously photographed it at Old Oak Common as D1723 in two-tone green livery.

Right: Built at Crewe in 1964, Class 47/4 **47459 (CD)** (D1579) started life at Gateshead, later working out of Thornaby, Stratford, Crewe Diesel, Holbeck (Leeds), York, Bescot and Sheffield (Tinsley) where it was unofficially named Perseus in 1991. Moving back to Crewe Diesel later that year, it was withdrawn in 1992 and, after storage, was finally broken up in 1993 by Booth Roe Metals, Rotherham. This image of 47459 heading a permanent way train is dated 14 May 1988. I would later photograph this loco at Leicester TMD on 10 March 1991.

One of my days off from the casino had coincided with the weekend and I had taken my son, Kevin, to London. I noted that we went to Hamleys, Harrods, Beatties, Holborn McDonalds and The Trocadero. At that time, I was still keeping a note of the diesels he saw and underlining them in a separate locoshed book in case, in the future, he ever wanted to know what locos he had seen; I remember wishing that my father had noted all of the steam engines I had seen as a child. However, I eventually had to accept that my passion was not my son's and the locoshed book remains incomplete.

Left: Built at Doncaster Works in 1979, **56070 (TO)** was first taken into stock at Toton where it stayed until 1993 when it was reallocated to Stewarts Lane; periods at Cardiff Canton and Immingham followed. After secure storage inside the closed Barry Wagon Repair Depot in 2000 followed by an examination at Toton, 56070 went to Thornaby in 2003. This image is dated 14 May 1988; I also photographed it at Stratford station in Transrail livery on 2 July 1999. It was cut up in 2011 by T. J. Thomson of Stockton. The underground line seen on the right-hand side of the signal box is the 'one stop' District Line from Earls Court.

Right: Brush built in 1966, Class 47/4 **47508 (OC)** *S.S. Great Britain* (D1952) was about to haul us on the 11.04 to Clapham Junction. It was always a Midland Region loco until 1973 when it took up residence at Landore, reallocation to Old Oak Common coming in 1977. Originally named Great Britain at Old Oak in 1979 it was renamed as above at Bristol Temple Meads station in 1985. The secondary plate below the nameplate reads: 'This locomotive was named by H.M. The Queen in 1985, the 150th anniversary year of the Great Western Railway'. 47508 kept this second name until 1992 when the nameplates were transferred to 47823. Further allocations were to be Crewe Diesel and Bristol Bath Road before withdrawal in 1993. It was stored at Bath Road in derelict condition until broken up on site by M. R. J. Phillips of Llanelly in 1995. This image is dated 14 May 1988; I had previously photographed 47508 at Liverpool Street on 3 January 1987 and this will appear in a future publication.

CLAPHAM JUNCTION STATION

Left: "Isn't it rich, are we a pair?" **33047 (SL)** and **33033 (EH)** hurry under the footbridge that spanned all of Clapham Junction's platforms. Both went new to Hither Green in 1961, both later worked out of Stewarts Lane and Eastleigh, both were withdrawn in 1993 and both were broken up at Eastleigh in 1997 by M. R. J. Phillips of Llanelly. Named Spitfire in 1991, 33047 was derailed at Hoo Junction in 1992 when its train ran away on the Grain branch and pushed it into a sand-drag, the rest of the train hitting the rear cab, causing severe damage. This image is dated 14 May 1988.

Right: Built by English Electric at Vulcan Foundry in 1966, **73119 (SL)** *Kentish Mercury* (E6025) alternated allocations between Stewarts Lane and Hither Green; it gained this name at Cannon Street station in 1986 and was rechristened Borough of Eastleigh in 2009, keeping it for two years; 73119 is still in service. The image is taken from the footbridge that spanned Clapham Junction station and every track seen here is a siding; not a through-line in sight. This image is dated 14 May 1988.

WATERLOO STATION

Right: Built by Birmingham Railway Carriage & Wagon Company, Class 33/0 **33008 (EH)** *Eastleigh* (D6508) went new to Hither Green in 1960, later working out of Stewarts Lane and Eastleigh. Sitting in the short parcels/loco bay on 14 May 1988, it surely looks eminently more stylish in its original livery of B.R. green than standard Rail Blue. 33008 is preserved at Battlefield Line Railway.

Left: 33008 is sitting next to another Class 33 destined for preservation, Class 33/1 **33103 (EH)** (D6514). Also built by BRC&WC of Smethwick, Birmingham, it went new to Hither Green in 1960, later allocations being Eastleigh and Stewarts Lane. Both this loco and 33113, pictured on the next page, were fitted for push-pull working and 33101-33119 were originally classified as Class 34. They were the mainstay of push-pull operations over the then unelectrified track from Bournemouth to Weymouth over three decades. Weymouth trains began at London Waterloo powered by third rail electric traction via Winchester and Southampton to Bournemouth where the train would divide, the 4REP remaining at the London end of the station and the 4-TCs hauled on to Poole and Weymouth by a Class 31/1. Returning, the locomotive propelled the train back to Bournemouth where it would be attached to a waiting London-bound 4REP and the locomotive detached to await the next Weymouth-bound portion. This image was taken on 14 May 1988 and 33103 is now preserved at the Ecclesbourne Valley Railway. Interestingly, although in private ownership at the time, 33103 and 4-TC 417 were spot-hired for use on Barking-Gospel Oak services during a stock shortage in 1999 and worked the service for some weeks without issue.

Left: Built by BRC&WC, Class 33/1 **33113 (EH)** (D6531) went new to Hither Green in 1960, later allocations being Eastleigh and Stewarts Lane. Withdrawn in 1992, it was cut up on site at Stewarts Lane by staff from M. R. J. Phillips of Llanelly. The Class 33/1 with one or two 4-TC sets (normally with the Class 33/1 at the country end of the train) were also the mainstay of the Waterloo-Salisbury service from their introduction. Platform congestion and the lack of facilities at the very busy Basingstoke station were two of the reasons for Class 33/1 operation throughout the route rather than just over the non-electrified section west of Basingstoke. Here, it is coupled to 4-TC 8027, British Rail Class 438, now preserved.

Right: Built by English Electric at Vulcan Foundry, **73136 (SL)** (E6043) went new to Stewarts Lane in 1966 and stayed until 1997, after which it alternated between Eastleigh and Hither Green. It would later be the recipient of two names, Kent Youth Music in 2002 and Mhairi in 2015; it remains in service. This image is dated 14 May 1988.

Built in 1968 by English Electric, Class 50 **50005 (LA)** *Collingwood* (D405) began life on the London Midland Western Lines, later being allocated to Crewe Diesel. Off to the Western Region in 1974, it then worked out of Bristol (Bath Road), Old Oak Common and Laira (Plymouth) where it gained the above name. Withdrawn in 1990, it was towed from Laira to Old Oak where, after being stripped of spare parts, it was broken up on site by Coopers Metals of Cardiff in 1991. This image is dated 14 May 1988. Waterloo-Exeter trains were worked by 'Warship' Class locos up until the 1971 mass withdrawal of Western Region diesel-hydraulics; after this, Class 33 diesels took over the service until Class 50s came over from the Western Region in 1979, after refurbishment at Doncaster.

NOTTINGHAM HOLDING SIDINGS

The date is 3 June 1989 and **20218 (TO)** (D8318) was the only Class 20 loco I hadn't seen, possibly because it went new to Haymarket (Edinburgh) in 1967 and only ever alternated home sheds between there and Eastfield (Glasgow) until a transfer to Toton in 1988. Even though Michael and I visited Scotland fairly regularly, 20218 had somehow slipped through the net. Other locos present on this date were 08416, 20026, 043, 069, 099, 136. Three months later, 20218 would be stored (unserviceable) at Toton, being withdrawn the following month after only a 22-year life. It was eventually cut up by M. C. Metal Processing, Springburn Works, Glasgow, in 1993. Surely, those wonderful factory buildings are an inspiration for modellers wishing to construct a realistic backscene. Obviously the bird would be a bit more difficult to replicate.

BLYTH CAMBOIS AND SURROUNDS

Crewe-built Class 56 **56120 (TO)** in large logo Rail Blue livery wends its way out of Blyth Power Station with a train of empty coal hopper wagons. Blyth Power Station, also known as Cambois Power Station, was a pair of now demolished coal-fired power stations, operating until 2001. The date is 19 August 1989 and, for the record, the Class 08 just managing to get into the picture is 08512, shedded at Blyth Cambois, which I 'copped'. And why have I not managed to incorporate the tops of the chimneys? Because the two on the right are 170m (560ft) high and the one on the left, not much less.

Left: This is the ill-fated Crewe-built Class 56 **56122 *Wilton-Coalpower* (TO)** in Railfreight Coal livery which, two years after this image was taken in the shed yard on 19 August 1989, would run through the buffers at Ryhope Grange Junction, near Sunderland, and suffer severe rear-cab damage when the following Hoppers piled into it. The damage was considered uneconomical to repair and it was withdrawn in 1992. It rotted away in the Training Compound at Toton until 1998 when it was moved to Booth Roe Metals, Rotherham and scrapped. New in 1983 to Sheffield (Tinsley), it also worked out of Gateshead and Toton. I photographed it inside Toton TMD in its damaged state in March 1992.

Right: Sitting in the yard at Blyth Cambois TMD is Crewe-built Class 56 **56126 (TO)** in large logo Rail Blue livery. New to Sheffield (Tinsley) in 1983, it later worked out of Gateshead, Toton and Immingham. Withdrawn in 1999, it was cut up the same year at Springs Branch (Wigan) CRDC. 'On shed' on 19 August 1989, were: 08421, 56113, 115, 118, 121, 122, 126, 127, 129, 131, 133.

After a satisfying visit to Blyth Cambois TMD I parked by the line at a wild spot on the Northumberland coast. Seeing a Class 37 approaching, I first saw the Motherwell Anvil and Hammer Depot plaque and I remember feeling satisfied that I had captured an unfamiliar loco with my camera. However, disappointingly, I was to discover later that Railfreight Coal liveried (37343) **37049 (ML) Imperial** (37322) (37049) (D6749) was very well known to me, as it had been based at Stratford four times in its life for a total of 15 years, in addition to spells at Norwich (Thorpe) and March. The date is 19 August 1989 and it seems strange that my son, who was sitting in the car with my wife, had seen this loco with me at Colchester TMD Open Day on 6 May 1985. But now, it is in Vera and Inspector George Gently country. You can almost imagine a dead body on the greensward surrounded by policemen and an actor, protesting in his best Geordie accent, "Why nor, Misser Ginley; ah doon't knoo oot aboot it".

CREWE AREA

The two with cast nameplates. Like the rest of the class, newly-built **92022 Charles Dickens** and **92023 Ravel** began life with their names in transfer form but were honoured with cast plates when, on 17 March 1995, they were chosen to double-head the first Class 92 scheduled freight through the Channel Tunnel. At Basford Hall open weekend on 27 August 1995, Class 92 92022 was positioned just inside the entrance to Basford Hall yard to greet us, quite a way from the other locomotives on show. 92022 meant something totally different to me, having obtained the smokebox numberplate of the 9F 'Crosti' steam loco years previously, but more of that in another book. Both Class 92s would both be allocated to Crewe Electric the following month. 92022 has since been exported to Bulgaria and 92023 was one of the Class 92s selected for working the Caledonian Sleeper between Scotland and London Euston.

Right: I am sure it will come as no big surprise to learn that Class 90/1 **90135 (CE)** *Crewe Basford Hall* was named at Basford Hall Rail Fair on 27 August 1995. Seen here in Railfreight Distribution livery, it had been new to Willesden in 1989 and was later allocated to Crewe Electric and Crewe Freightliner. This loco remains operational.

Left: Crewe-built Class 47/0 **47053 (TI)** *Dollands Moor International* (D1635) was allocated new to various divisions of the LMWL in 1964, later working out of Immingham, Sheffield (Tinsley), Inverness, Haymarket (Edinburgh), Eastfield (Glasgow), Bescot, Cardiff Canton and Crewe Diesel, ending up at Tinsley for a second time in 1990 where it gained an unofficial name, Impala. It was later officially named Cory Brothers 1842-1992 until the nameplates were removed after two years when it acquired the above name. Withdrawn in 1999, it then became a long-term inmate of Barrow Hill until in 2007 it was taken by road to European Metal Recycling, Kingsbury, and broken up that year. This image of 47053 in Railfreight Distribution livery is dated 27 August 1995.

The date is 7 July 1991 and Class 87/1 **87101 (CE)** *Stephenson* had been derailed and damaged the previous March at Lichfield Trent Valley station and still appears to be out of action, sitting at Crewe station in the line of locos awaiting Works attention. It would later pull Michael and me twice on 3 May 1997, from Euston to Crewe Electric TMD Open Day and return. That was a good trip as we saw 23 of the 'new' Class 92s; add to this the 14 we had already seen at the Basford Hall Rail Fair and it didn't leave many of the fleet of 46 left to 'cop', a task which I have now completed. As can be seen in the inset image, one of the nameplates has pride of place outside the public toilets at Shildon Railway Museum. No doubt flushed with the success of securing this item, the Shildon management made an executive decision to display it for the convenience of visitors. Obviously, the museum at Shildon wasn't renamed 'Locomotion' for nothing! The date is 29 October 2012 and, joking aside, the magnificent experience at Shildon definitely makes it worth a visit.

New nameplates were fitted in 1994 and on 27 August 1995, Class 87/1 **87101 (CE)** *Stephenson* displays them at Basford Hall. The name had originally belonged to 87001, which then became Royal Scot. Originally intended to be numbered 87036, this loco was given a thyristor power control system which improved its hauling ability by 20%, thus making it more suitable for freight traffic. Just a couple of notes I have of seeing it travelling through Stratford station were on 5 February and 17 September 1998, both light-engine. Withdrawn in 1999, it was used as a source of spare parts and was totally scrapped by 2002.

Right: **20075 (BS)** *Sir William Cooke* (D8075) went new to Eastfield (Glasgow) in 1961, later allocations being Thornton Junction, Polmadie, Toton, Bescot, Sellafield Servicing Depot and Carlisle (Kingmoor). It was one of four Class 20s allocated to BRT (BR Telecommunications Ltd) with Telecom-related names but none of them did very much revenue earning work. The livery hardly needs description in detail; the date is 27 August 1995 and 20075 is still operational.

Left: On 27 August 1995 Brush-built Class 31/1 **31105 (BS)** *Bescot TMD* (D5523) was 36 years old, going new to Stratford in 1959. It later worked out of Ipswich, March, Immingham, Bescot and Thornaby. The additional plate below the nameplate actually reads ***Bescot & Saltley Quality Assured*** but, for some reason, '*& Saltley*' has been erased. A number of diesel locos were named after Depots and they all had 'Quality Assured' plates accompanying them. I am pleased to report that this loco is preserved and currently resides close to me at Mangapps Railway Museum, Burnham-On-Crouch; the slight downside is that it carries high-visibility yellow livery; it is difficult to imagine a more unflattering contrast to the wonderful B.R. green it wore when allocated new to Stratford. I saw it recently, not realising at the time that it was the same '31' I photographed in Transrail colours at Basford Hall Open Day. It is now almost 65 years old.

And here it is, the first-ever Class 47, Brush-built **47401 *North Eastern*** (D1500), withdrawn in 1992 and now preserved. New to Finsbury Park in 1962, it later worked out of Sheffield (Darnall), Sheffield (Tinsley), Immingham and Gateshead. It gained the above name in 1981 but lost it again in 1988. Later unofficially named Star of the East, it was withdrawn in 1992 and sold for preservation to the 47401 Project where, as can be seen, it regained its old name. The date is 27 August 1995.

I am trying very hard to resist saying, 'an exceedingly good loco', but I appear to have lost the fight; sorry, the joke had to be made, but I'm sure everybody knows that **92034 *Kipling*** was named after one of this country's most influential authors and not a 'Country Slice'. On this date, 27 August 1995, 92034 has not yet been accepted into stock.

The last day of service of **'The European'** (Harwich Parkeston Quay-Edinburgh and Glasgow) was 14 May 1988, hauled by **86604 (CE)** (86404) (86004) (E3103); I travelled behind it from Shenfield as far as Watford Junction; it is the loco that rushed past me at the Margaretting lay-by, a mile from my home on the previous day to that but I'm sure I had no idea of this when I clicked the shutter here at Basford Hall on 27 August 1995; it had changed in appearance so much with a new number and livery. The story, and details of the locomotive, can be found earlier in this book.

Left: I photographed this loco from both sides 'to be sure, to be sure' as the old joke goes, but yes, the name was spelled incorrectly on both sides; I wonder how long it took the powers that be to discover this. On this date, 27 August 1995, **92016 Brahms** has not yet been accepted into stock.

Right: A fellow author? I should be so lucky (once more unto the Depot, dear friends, once more). On 27 August 1995 **92017 Shakespeare** has not yet been accepted into stock. There were 14 of the new Class 92s at Basford Hall on this day, thus giving me a good base upon which to build in my quest to see the whole class. I am pleased to say, that objective was achieved some time ago.

Class 90/1 (90028) **90128 (CE)** *Vrachtverbinding* (90028) in Belgian (SNCB) livery at the Basford Hall Rail Fair on 27 August 1995. New to Willesden in 1989, it was later allocated to Crewe Electric and Crewe Freightliner. In 1992 it had been decided that three Class 90s would receive the application of three European railway colours as part of the Freightconnection exhibition held at the NEC, Birmingham. The UK Class 90 named Freightconnection was not present at Basford Hall Open Day but I had previously photographed it at Kings Cross station and it appears elsewhere in this book. 90128 is currently in store at Crewe Electric Depot.

Class 90/1 (90029) **90129 (CE)** *Frachtverbindungen* (90029) in German (DB) colours at the Basford Hall Rail Fair on 27 August 1995. New to Willesden in 1989, it was later allocated to Crewe Electric and Crewe Freightliner. This name was incorrectly spelt on the first pair of nameplates fitted (not difficult to do, one would imagine) and new nameplates were cast. This loco is currently in store at Crewe Electric Depot.

Class 90/1 (90030) **90130 (CE)** *Fretconnection* (90030) in French (SNCF) colours at the Basford Hall Rail Fair on 27 August 1995. New to Willesden in 1989, it was later allocated to Crewe Electric and Crewe Freightliner. These nameplates were removed in 2000 and replaced by Crewe Locomotive Works. This loco is currently in store at Crewe Electric Depot.

A loco that fooled me twice. I photographed English Electric-built Class 37/5 **37517 (TE)** *St. Aiden's C. E. Memorial School, Hartlepool, Railsafe Trophy Winners 1995* (37018) (D6718) at Basford Hall Rail Fair on 27 August 1995 and, when compiling this narrative, I wondered what it looked like prior to carrying Loadhaul's distinctive colours and a very large nameplate. It transpired that it was a Thornaby stalwart (see inset) I had seen many times wearing Railfreight Red Stripe livery. I then looked at the notes I am compiling on Thornaby and see that this was 'déjà vu, all over again' as the joke goes. This is D6718 'as was' (to use an expression my maternal grandmother employed when referring to a married woman by her surname prior to marriage). In other words, a loco I had seen it many times in B.R. green on the ex-G.E. It was purchased by West Coast Railways and therefore now boasts a more subdued livery. Not a lot of people see it, though, as it is currently stored.

Left: Crewe-built **47972 (CD)** *The Royal Army Ordnance Corps* (97545) (47545) (D1646) went new to Cardiff Canton in 1964, future allocations being Landore (Swansea), Bristol (Bath Road), Bescot, and Crewe Diesel where it was named The Royal Army Ordnance Corps. It went on to work out of Immingham and was broken up by C. F. Booth of Rotherham in 2010. I had previously photographed it at Leicester TMD on 24 May 1992 in large logo Rail Blue livery and that image appears elsewhere in this book. Here, it displays a one-off livery for a '47' of a dark pink upper bodyside that extended around the cab windows with a half-yellow warning end. A 'technical services' badge was applied to the mid-grey lower body.

Right: Crewe-built **47973 (CD)** *Derby Evening Telegraph* (97561) (47561) (47034) (D1614) went new to Worcester in 1964, future allocations being Cardiff Canton, Old Oak Common, Landore (Swansea), Bristol (Bath Road), Bescot, and Crewe Diesel where it was named Midland Counties Railway 150 1839-1989 in May 1989; I have a photograph as it as such in St. Pancras station on 19 July 1989 carrying Midland Counties dark red livery as opposed to the InterCity Executive or Main Line livery it carries here. In 1990, it was given the name it carries in this image dated 27 August 1995. Withdrawn in 1996, it was cut up on site at Crewe Works by M. R. J. Phillips of Llanelly in 1997.

It is difficult to believe that this loco is the same one that appears as the opening photograph in Stratford Depot Locomotives in The Eighties and Nineties when it was named Rail Riders and, obviously, I never realised this whilst taking this shot. Brush-built **47488 (CF)** *Davies The Ocean* (*Waterman Railways* plate above it and **Heritage Class** plate below it) (D1713) went new to Cardiff Canton in 1964, then Bristol (Bath Road), various divisions of the Midland Region, Bescot, Crewe Diesel and, ending up where it had begun life, Cardiff Canton. Withdrawn in 2003, it is currently stored. The two men demonstrating the skill of rail-joining have not attracted an enormous crowd but they appear to be enjoying themselves.

This image personifies what happened to so many of the Class 47 locos; they were either involved in a collision or they caught fire. I would like to see the statistics; indeed, I may compile them myself. Class 47/4 **47850 (CD)** (47648) (47151) (D1744) was damaged in an accident at Crewe station a couple of months prior to this image being taken on 27 August 1995; it was moved to Crewe Works the following month for assessment and was withdrawn from service. Remaining at the Works for another two years, it was cut up on site by M. R. J. Phillips of Llanelly in 1997. It had gone new to Cardiff Canton in 1964, other allocations being Old Oak Common, Bristol (Bath Road), Bescot, Landore (Swansea), Laira (Plymouth), Carlisle (Kingmoor) and Crewe Diesel. I have a note that I photographed this loco in Paddington station on 20 December 1992.

Class 31/1 **31255 (SP)** (D5683) went new to Sheffield (Darnall) in 1961, future allocations being Finsbury Park, March, Sheffield (Tinsley), Stratford, Bescot, Crewe Diesel, Immingham and Springs Branch (Wigan). Withdrawn in 1999, it was sold to The Colne Valley Diesel Group, Castle Hedingham; it remains in preservation on The Mid-Norfolk Railway. This image of 31255 in Civil Engineers 'Dutch' livery is dated 27 August 1995.

BIRKENHEAD NORTH TMD

Left: In all-over BR Rail Blue livery with black buffer beams and yellow and black 'wasp' warning ends, Darlington built Class 08 **08532 (AN)** (D3694) had been an Allerton loco for 17 years. New to Sheffield (Darnall) in 1959, it had only resided at Gorton, Reddish and Longsight (Manchester) before settling at Allerton in 1973; it was on shed when I visited Allerton on 3 February 1990. Withdrawn in April 1993, 08532 was stored at Allerton until broken up on site by Coopers Metals of Sheffield early the following year. This image was taken on 15 July 1990 when 08532 was one of only three locomotives on shed at Birkenhead North, the other two being battery locomotives. The word 'corrosive' can be discerned on the sticker, but nothing else.

Right: **DB 977362** (ex 97701) was one of two Departmental battery locos [the other was DB 977363 (ex 97702)] that were converted from Euston-Watford Class 501 EMUs. The Driving Motor Brake Standards were used; in other words, the end coaches of the original sets. In these cases, the original Class 501 units were M61136 and M61139 respectively and were converted at Wolverton Works in May 1974. As well as battery power, these were the only two battery locos that also retained their third rail pick up shoes for working on the Merseyrail DC system. They remained at Birkenhead until scrapped in 1993. Strange to think that Michael and I may have travelled from Euston to Willesden Junction in one of this pair of coaches.

CARLISLE CITADEL STATION

This is my favourite shot of a Class 08 and I'm not sure why. Maybe because the station structure, the wet platform and the BR General all-over mid-grey livery seem to complement each other. If you are unfamiliar with the following shedcode, it had somehow escaped me, too. It is Carlisle Upperby Carriage Depot. Darlington-built in 1959, **08534 (CL)** (D3696) went new to Sheffield (Darnall), future allocations being Annesley, Toton, Kirkby-in-Ashfield, Burton-on-Trent, Longsight (Manchester), Springs Branch (Wigan) and Allerton, coming to Carlisle in 1988. Withdrawn in 2002, it was cut up by T. J. Thomson and Son, Stockton. This image is dated 19 August 1991.

Right: The date is 28 May 1984 and I have a note of seeing this loco in BR green as D5401 in Cricklewood Diesel Depot on 3 July 1965, along with D5398, D5400, D5402 and D5409. I couldn't have guessed at the time that, within two or three years, these Type 2 Birmingham R. C. & W. Co. locos would have emigrated across the border. Originally, the Class 27s were used on push-pull rakes of Mark 2 coaches between Edinburgh and Glasgow, a Class 27 at each end. A long-term Eastfield (Glasgow) resident, **27056 (ED)** (27112) (D5401) awaits the 'Right Away' from Carlisle Citadel. Withdrawn in 1987, this loco is now preserved on the Great Central Railway after rescue from Vic Berry's scrapyard.

Left: Weaving its way through Carlisle Citadel station on 19 August 1991, (90038) (90238) (90038) **90138 (CE)** (90038) carries standard triple-grey Railfreight Distribution livery. New to Willesden in 1989 but now at Crewe Electric, it will be later allocated to Crewe Freightliner.

37902 (CF) *British Steel Llanwern* (37148) (D6848) was converted to a Class 37/9 in 1986; new bogies were fitted and it had ballast weights to increase the overall weight to 120 tons. The Railfreight triple-grey Metals livery used downward facing chevrons, supposedly representing steel girders; a bit of a stretch, some would say. New to Cardiff Canton in 1963, it later worked out of Haymarket (Edinburgh), Eastfield (Glasgow), Motherwell, Landore (Swansea), Bristol (Bath Road) and March. It gained the above name in 1991 but only carried it for just over a year. Withdrawn in 1998, it was cut up by Sims Metals in 2005. It is seen here with the famous detached West Wall of Carlisle Citadel station as a backdrop on 19 August 1991. The wall is a Grade 2 listed building; not many people know that!

CARLISLE UPPERBY TMD

Left: Class 31/1 **31270 (CD)** (D5800) is a loco I remember well from the ex-G.E. when it carried B.R. green livery and a running number of D5800, but how could I have known when taking this photograph? New to March in 1961, further allocations were Leeds (Holbeck), York, Sheffield (Tinsley), Immingham, Toton, Carlisle (Kingmoor), Thornaby, Crewe Diesel and Springs Branch (Wigan). Initially sold to The Colne Valley Railway in 2000, it remains in preservation and has been the recipient of at least two names that it never carried in real life. This image of 31270 in Railfreight triple-grey livery with Coal Sector markings is dated 19 August 1991.

Right: Derby-built Class 08 **08808 (WDN)** (D3976) had gone new to Cardiff Canton in 1960. Further allocations were Oxley (Wolverhampton), Stourbridge Junction, Bletchley, Carlisle (Kingmoor) and Carlisle Upperby Carriage Depot. Withdrawn in 1990 and having had all reusable parts removed, as can be seen, the body shell was dumped in Carlisle (Upperby) car park with a view to cutting it up on site. However, moved by road to Booth-Roe Metals, Rotherham in 1992, it was scrapped by them that year. This image is dated 19 August 1991. Princess Margaret Rose keeps a watchful eye.

EXETER ST DAVID'S STATION

Right: Built by English Electric, **50007 (LA)** *Sir Edward Elgar* (D407) was reallocated from Crewe Diesel to Laira (Plymouth) in 1974. When the Class 50s gained their names in 1978, they were named after Royal Naval ships with notable records in the two world wars including 50007 which was originally named Hercules. However, in 1984, to celebrate the 150th anniversary of the Great Western Railway and the 50th anniversary of the composer's death, it was repainted in lined Brunswick Green livery, given cast brass numberplates, a GWR brass crest, cast double-arrows and it gained the name it carries here. Withdrawn in 1991, it was reinstated for a Class 50 rail tour and withdrawn a second time in 1994; it is now preserved. This image is dated 2 August 1989.

READING TMD

Left: Crewe-built Class 47/4 **47484 (OC)** *Isambard Kingdom Brunel* (D1662) went new to Landore (Swansea) and within one month had been given the name it still carries here, albeit a different casting. Later allocations were Cardiff Canton, Bristol (Bath Road) and Old Oak Common in 1985 where it received its second set of nameplates. It then spent an inordinate amount of time in store at various locations, occasionally being reinstated; in 2011 it was sold to the Pioneer Diesel Loco Group. This image is dated 3 November 1990.

ST. BLAZEY TMD

St. Blazey is located in Par, on the main line a couple of miles west of St. Austell, Cornwall; the shed closed in 1987, with diesel locos being serviced at the Wagon Repair Depot. As I was with my son, Kevin, I asked permission to take some photographs and we were shown around by a driver. He lived locally at a place I had never heard of until then, Burngullow; thank goodness he had introduced himself as Ted because, had his name been Bill, I doubt I could have resisted embarrassing my son by launching into, 'Hey, Burngullow Bill'. Locos stabled at St. Blazey were almost exclusively engaged in the transport of china clay by rail.

Right: English Electric Class 37/5 **37669 (LA)** (37129) (D6829) would later be badly damaged in a collision with 37411 at Burngullow Junction in 1993 (yes, where Ted lived, but the collision had nothing to do with him, as far as I know!) It began life at Cardiff Canton in 1963, future allocations being York, Thornaby, Healey Mills, Gateshead, Sheffield (Tinsley), Stratford, Motherwell, Bristol (Bath Road), Inverness and Eastfield (Glasgow) before arriving at Laira (Plymouth) in 1988. After repair at Crewe Works in 1995, more storage ensued; its last two allocations were Toton and Crewe Diesel. Later purchased by West Coast Railways, this image of 37669 carrying Railfreight large logo Red Stripe livery is dated 30 July 1989.

Like the other Class 37s pictured here at St. Blazey on 30 July 1989, **37412 (LA)** (37301) (D6601) is another Laira-based loco engaged in china clay traffic. It had previously been allocated to Inverness and, previous to that, Eastfield (Glasgow) where it had been named Loch Lomond; it can clearly be seen where the nameplate used to be and, for some reason, 37412 still carries a 'Scotty Dog' logo on its standard large logo Rail Blue livery. It was allocated to Landore (Swansea) in 1965 when new, moving to Cardiff Canton after a couple of years and in 1994 it moved back to Canton (and was named Driver John Elliott) after this stint at Laira (Plymouth). Later working out of Toton and Crewe Diesel, 37412 was scrapped at European Metal Recycling, Kingsbury, in 2012.

Part of the distinctive St. Blazey (83E) steam roundhouse engine shed roof is visible to the right of Robert Stephenson & Hawthorns-built Class 37/5 **37675 (LA)** *William Cookworthy* (37164) (D6864) on 30 July 1989. It comprised nine 70ft roads around a turntable which is seen here to the left of the loco. Although the roundhouse has since been converted into industrial units, the turntable has been retained to turn the preserved steam locomotives that still visit Cornwall on special Main Line workings. In Railfreight large logo Red Stripe livery, it is another Laira-based loco working china clay traffic duties. The loco's name is appropriate for this part of the country, William Cookworthy being the first person in Britain to discover how to make hard-paste porcelain, similar to that imported from China. He was also an English Quaker minister, so I think it is safe to say that he certainly knew his oats!

MARGARETTING

Left: Built by B.R. at Doncaster Works in 1965, Class 86/2 **86221 (WN)** *BBC Look East* (E3132) passes Margaretting on 22 May 1987 in InterCity Executive colours heading a train similarly liveried. Allocations were, as usual, 5F, ACL and Willesden, and it had been named Vesta between 1979 and 1987. Gaining the above name at Liverpool Street station in 1987, further allocations were Ilford and Norwich (Crown Point). Stored at Springburn Works (Glasgow) in 2003, it was taken by road to Immingham Railfreight Terminals that year and scrapped by staff working for Easco. I note that I saw this loco double-heading with 47537 (BR) Sir Gwynedd – County of Gwynedd (leading) into Liverpool Street on 26 May 1988, the assumption being that 86221 had failed.

Right: Doncaster-built Class 86/4 (86604) **86404 (WN)** (86004) (E3103) takes the last eastbound **'European'** past my favourite ex-G.E. vantage point at Margaretting on 13 May 1988, my trusty Ceramic Blue Mark 1 Sierra also featuring. I photographed 86404 at Shenfield station the next day on the return journey before travelling on that train as far as Watford Junction. That image and more details of the last **'European'** can be found under 'Shenfield station' in this book. I also photographed it on 27 August 1995 at Crewe Basford Hall in another incarnation as 86604 in Railfreight Distribution livery; that image is also contained in this book. 86404 went new to 5H in 1965, followed by ACL, Willesden, Crewe Electric and Crewe International Electric (Freightliner) in 1998. This loco is still in service.

CHELMSFORD STATION

A 'parcels' arrives at Chelmsford station behind **47291 (SF)** *The Port of Felixstowe* on 20 March 1987. There used to be a passing line between these two tracks but this has been removed; luckily, the five-storey high signal box, unused since 1994, still survives. Sidings at the eastern end of the station used to serve a goods yard close to Hoffman's ball-bearing factory; these sidings were reduced to serve a mail sorting office and a building materials yard, the former long-gone but the latter still in use and I often see a Class 66 loco depositing and shunting wagons. Now converted into living accommodation, my son, Kevin, lives in the former ball-bearing factory where, many years ago, my late cousin Richard used to work. I really must mention Richard; also a railway enthusiast, he is the only person I know who died three times on three separate occasions. He was older than me and, as a boy living in Grays on the ex-LT&S, I remember thinking it was amazing that Richard would make the 27-mile journey from Chelmsford on a motorcycle. Not only that, he made the journey in a remarkably short time and I remember my Aunty Joan saying, "He'll kill himself on that thing"; and he very nearly did, but not quite. Still a teenager, he had a serious accident in Chelmsford and was pronounced dead at the scene. He was taken to the morgue but, luckily, an attendant noticed a toe twitching and Richard was rushed up to a ward. He was hospitalised for six months but managed to stay alive for another 50 years until a routine visit to the dentist became another opportunity for a close encounter with the Grim Reaper; he had a heart attack in the dentists. His heart stopping beating for a short while but, thanks to a defibrillator that all dentists carry as standard (a bit worrying, that dentists feel they need such an item) he survived again. A few years later, sadly, he died for real and the oracle couldn't be worked a third time. I remember saying to his wife after the funeral, "I can't believe he won't come walking back through the door". I know everybody says that in such situations, but I actually meant it.

On 3 June 1995, Class 37/0 **37023 (SF)** *Stratford TMD Quality Approved* (D6723) is photographed in Mainline Blue entering Chelmsford station on the outward journey of a Hertfordshire Rail Tours excursion entitled 'The Pheasant Plucker'. Had there been an over-subscription and they needed to put on a second train, I wonder if it would have been called 'The Pheasant Plucker's Son'? (The route was London Liverpool Street–Ipswich–Lowestoft–Norwich–Ely–Cambridge–Bishops Stortford up loop–Broxbourne up loop–Cheshunt–Turkey Street–London Liverpool Street). New in 1961, this English Electric-built loco alternated home Depots between March and Stratford for 21 years and so had obviously passed through Chelmsford thousands of times in B.R. green until it was reallocated to Healey Mills in 1982; Thornaby, Sheffield (Tinsley), Motherwell and Eastfield (Glasgow) followed. In 1991, 30 years after hauling Liverpool Street to Norwich expresses, 37023 would be transferred back to Stratford where it gained the name Stratford. Reallocations to Toton and Motherwell followed until 37023 went into long-term storage at Old Oak Common. Withdrawn in 2009, it was purchased privately and is currently undergoing restoration at the Pontypool & Blaenavon Railway. I photographed 37023 (unnamed) in Thornaby TMD on 3 June 1989 in standard B.R. Rail Blue and at Ipswich refuelling point in 'Dutch' livery in 1993 as Stratford, prior to it gaining the TMD Quality Approved add-on in 1994.

Above: The late Queen Elizabeth and Duke of Edinburgh came to Chelmsford by train on 29 July 1988 and I was curious as to what loco would haul the dedicated set of claret carriages, not to mention see the Queen walk along the same platform as us commuters. The Royal Train was hauled by Class 86/2 (86901) **86253 (WN)** *The Manchester Guardian* (86044) (E3136), one of the many hundreds of locomotives immaculately prepared by Stratford to haul Royal trains over the years. Built by B.R. at Doncaster Works in 1965, it went the usual route of 5H, ACL and Willesden, gaining the above name at Manchester Piccadilly station in 1980. Unfortunately, 86253 was destined to catch fire and suffer serious damage in 1997 whilst working a Euston to Glasgow express. After repairs at Glasgow Springburn Works, it came back 'down south' in the consist of a Mossend to Willesden freight hauled by 90021. Future allocations were Norwich (Crown Point), Willesden and Longsight (Manchester). It was later painted yellow when it became 86901 Chief Engineer, the 'Master' in a 'Master & Slave' combination with another Class 86 loco in a conversion to a Mobile Load Bank. It was scrapped at Sandbach Commercial Dismantlers in 2018.

Left: The driver looks worried. "Not this side, Your Majesty!"

EDINBURGH WAVERLEY STATION

Edinburgh Waverley railway station is the principal station serving Edinburgh, Scotland, and is the second busiest in Scotland after Glasgow Central. It is the northern terminus of the East Coast Main Line, 393 miles 13 chains (632.7km) from London Kings Cross and is situated in a steep, narrow valley between the medieval Old Town and the 18th century New Town; Princes Street, the premier shopping street, runs close to its north side. In the early days, Michael and I slept on hard benches in the waiting room when funds were running short. I remember the smell from a nearby brewery was more noticeable at night and almost overwhelming.

Left: Class 47/4 **47550 (ED)** *University of Dundee* (D1731) carries yet another variation of Executive InterCity livery inasmuch as the yellow warning ends are not 'wraparound', compared, for example, to 47643 in the Inverness section of this book. It began its working life at Old Oak Common where it stayed for three years before transferring to Cardiff Canton and I note I saw it on frequent occasions at Old Oak Common and Paddington. Then came a spell on various divisions of the London Midland Region and Bescot before it made its way back to Scotland with allocations to Eastfield (Glasgow) and Inverness; it was named at Dundee Shore Road sidings in 1982. Staying in Scotland until 1991, it then went into store at Crewe Diesel, Immingham TMD and Immingham Railfreight Terminals, the nameplates being removed and 47550 authorised for component recovery. In 2008 it was removed by road to European Metal Recycling and broken up in 2010. This image is dated 19 August 1989.

Right: Brush-built Class 47/7 **47701 (ED)** *Saint Andrew* (47493) (D1932) began life at Bristol (Bath Road) in 1966, moving to Landore (Swansea) in 1974 and arriving at Haymarket (Edinburgh) in 1977 where it gained the above name. It then collided with a reversing freight train at Winchburgh Junction and suffered severe damage to its cab and was moved into a siding at Dalmeny, by the Forth Bridge. Eventually repaired, it was reallocated to Old Oak Common where it was named Old Oak Common Traction & Rolling Stock Depot, further allocations being Eastleigh and Crewe Diesel. The nameplates were removed in 1993 and passed on to 47004, which features in the Eastfield section of this book. Not wishing to remain nameless, it then became Waverley but misfortune befell it again when it received serious underframe and bogie damage at Birmingham New Street station. Repaired yet again, it was purchased by Nemesis Rail and is currently stored 'non-operational'. This image of 47701 in a variation of InterCity ScotRail livery is dated 19 August 1989.

Brush-built Class 47/7 **47715 (ED)** *Haymarket* (47502) (D1945) was new in 1966 and worked on various divisions of the London Midland Region until 1972 when it moved to Bristol (Bath Road), Landore (Swansea) and Old Oak Common before being reallocated to Haymarket (Edinburgh) in 1985 where it gained the above name. Future allocations were Eastfield (Glasgow), Old Oak Common, Eastleigh, Stratford and Crewe Diesel. Withdrawn in 1999, it was preserved and named Poseidon and is currently stored 'non-operational'. This image of 47715 in a variation of InterCity ScotRail livery is dated 19 August 1989.

EUSTON STATION

A familiar backdrop; the 'Mobil' sign gives away the fact that this is Euston. With only six months before withdrawal due to collision damage, Brush Class 31/1 **31109 (BS)** (D5527) is on borrowed time at Euston station on 24 September 1987, hauling the test-coach 'Mentor'. In 1973, a new overhead line test coach was introduced which bore the name 'Mentor'. This was an acronym for 'Mobile Electrical Network Testing, Observation and Recording'. The Mentor coach was converted from a B.R. Mk1 BSK coach and extensively rebuilt to provide an area for the pantograph as well as an observation position and was numbered ADB975091. After the electrification between Weaver Junction and Glasgow was completed in early 1974, the Queen and the Duke of Edinburgh rode on the vehicle which was marshalled into a small version of the Royal Train on 7 May 1974; the coach was still in use with Network Rail in 2020. 31109 carries all-over Rail Blue with yellow warning ends and is a 'skinhead' (no headcode box). Prior to withdrawal in March 1988, 31109 had been a much-travelled locomotive, beginning at Stratford when new in 1959. It later spent time at Ipswich, Sheffield (Darnall), Wath, Lincoln, Immingham, March, Gateshead, Bescot and Crewe Diesel, ending up full circle at Stratford again in January 1988, two months prior to a collision that ended its career. Images of 31109 at Stratford after the collision appear in another book in this series, Stratford Depot Locomotives in The Eighties and Nineties.

Left: Brush-built Class 47/7 **47776 (CD) Respected** (47578) (47181) (D1776) enters Euston on 27 August 1995 with the empty stock that would convey Michael and me to Crewe. New to Sheffield (Tinsley) in 1964, future allocations were Finsbury Park, Stratford, Leeds (Holbeck), Immingham, going to Eastfield (Glasgow) in 1980 where it was renumbered 47578 and named The Royal Society of Edinburgh. If this is beginning to sound familiar, this loco is pictured as 47578 in both the Eastfield and Inverness sections of this book. Withdrawn in 1991, it was reinstated the following year to Crewe Diesel, given B.R. parcels livery and named Respected. It is currently stored 'non-operational' for West Coast Railways.

Right: Class 56 **56049 (CF)** went new to Toton in 1978, later working out of Cardiff Canton, Bristol (Bath Road), Stewarts Lane and Immingham. In 1998 it suffered severe cab damage when it hit a bogie bolster wagon in Warrington Yard. Repaired at Toton, but not until 2001, it was then allocated to Thornaby. It later worked in France and, on return to the UK was sold to Colas Rail. It is still operational and, in 2018, gained the name Robin of Templecombe 1938-2013. This image is dated 27 August 1995, just prior to it hauling us to Crewe.

Doncaster-built Class 58 **58015 (TO)** enters Euston acting as a 'Thunderbird' on 26 February 1990 when overhead power lines had come down further north, thus necessitating that diesel locomotives came to the rescue. 58015 had been delivered new to Toton in 1984, its only UK allocation, but it later worked in France and Spain. Stored at Monforte-del-Cid, Spain, in 2016, it was eventually scrapped in 2019.

Above: 58015 now sits next to Brush-built Class 47/4 (47771) **47503 (CD)** *The Geordie* (D1946) doing the same job. New in 1966, 47503 worked on various divisions of the LMWL until 1973, later allocations being Bescot, Crewe Diesel, Carlisle (Kingmoor) and Sheffield (Tinsley). Losing this name in 1991, it became Heaton Traincare Depot in 1993. Sold to a private purchaser, it is currently in store. Interesting as this whole situation was, I confess that I was more intrigued by one of the cars parked on the platform. Sometimes, the most amazing of coincidences seem to happen to me and this was one of those times. My attention was drawn, not to the Vauxhall Senator on the left but to the Ford Granada Estate on the right. Amazingly, its registration of C333 FHK is an amalgamation of my father's first two cars. He passed his test late in life and his first two vehicles, both Ford Cortina Mark Ones, were **333** FLX (which the family christened 'Felix') and **FHK** 969B. And when you look at the red nameplate on the '47' and take into account that my father's name was George, well! As my paternal grandmother would have said, you could have knocked me down with a feather. The date is 26 February 1990.

Left: Doncaster-built Class 58 **58012 (TO)** has assisted Crewe-built Class 90 mixed traffic loco **90030 (WN)** (90130) (90030) which has apparently been 'caught short'. 58012 had been delivered new to Toton in 1984, its only UK allocation. Put into store in 1999, it was heavily stripped of parts but, despite that, was purchased for preservation in 2016 by the Battlefield Line and awaits restoration.

A different view of the pair. 90030 had been taken into stock at Willesden in May 1989, but did not enter revenue earning service until November 1989. The date here is 26 February 1990 and 90030 would be reallocated to Crewe Diesel the following month. It was later named Fretconnection in 1992 and Crewe Locomotive Works in 2000. It can be seen in French (SNCF) livery at Crewe Basford Hall earlier in this book and is currently in store. Looking back, I should have stayed around at Euston that day to see what other unusual diesel locos put in an appearance but, as in the early days at sheds or stations, the practice of 'note what was there and move on to the next location' seemed embedded in my psyche.

Class 82 **82008 (WN)** (E3054) was built by Beyer Peacock Ltd. in 1961 and is pictured at Euston on 14 March 1987, nine months prior to final withdrawal. It had previously been withdrawn in 1983, along with the rest of the Class, but was reprieved to work between Euston and Willesden on empty coaching stock duty. Painted in InterCity livery, it has a lower positioned silver coloured metal InterCity arrow instead of a higher painted white one. The Class 82 originally comprised ten locomotives but, as the first and last (E3046 and E3055) were destroyed by fire, only the remaining eight were eligible for a TOPS number. E3046 hauled Michael and me from Rugby to Crewe on 27 August 1965 and was withdrawn in 1971. I have a photograph of it outside Crewe Works covered in a tarpaulin after the fire and this image will appear in another book. 82008 had been allocated to 5H, ACL, Longsight Electric Depot (Manchester) and, finally, Willesden. It is now preserved at Barrow Hill Roundhouse Museum.

FORT WILLIAM STABLING POINT

English Electric built **37405 (ED)** *Strathclyde Region* (37282) (D6982) was allocated to Cardiff Canton when new in 1965 and stayed until moving to Eastfield (Glasgow) in 1984. Named at Glasgow Queen Street station in 1986, it was in collision with DMU number 156456 in 1991 and sustained severe end damage. 37405 was moved to Eastfield for assessment and then to St. Rollox Works (Glasgow) for repair, returning to service at Motherwell the following year. Future allocations were Sheffield (Tinsley), Immingham, Crewe Diesel, Springs Branch (Wigan) and Stratford, eventually arriving at Toton in 1997 where the nameplates were removed. In 2007 it worked on the upgrade of the Manchester Metrolink on ballast workings and, after storage, it was sold to Direct Rail Services in 2012. Still retaining snowploughs, it was one of a batch of Direct Rail Class 37s sent to Norwich to top and tail workings to Lowestoft and Great Yarmouth and, on 14 May 2019, topping and tailing with 37716, it hauled me and three friends, Michael Fanthorpe, Tim Mercer and Phil Starling from Norwich to both of those destinations and 'tailed' us on the return journeys. As I photographed it in its Direct Line livery, how could I have possibly realised that I had photographed it 30 years previously as Strathclyde Region at Fort William stabling point on 21 August 1989? As can be seen, back then it carried standard B.R. Rail Blue livery with wrap-around yellow warning ends, large double arrow and a large West Highland Terrier logo; 37405 is still operational. I hope somebody recognises the lad taking a video out of the carriage window, and tells him.

English Electric built **37409 (ED)** *Loch Awe* (37270) (D6970) was allocated to Cardiff Canton when new in 1965 so this is surely only one of many times it has sat in close proximity to 37405. Landore (Swansea) came next, followed by Laira (Plymouth), Eastfield, Motherwell and then a period of storage. Named Lord Hinton in 2010, 37409 is still operational for Direct Rail Services, along with 37405, still together after all these years. Carrying InterCity Executive livery with yellow warning ends, the date of this image is 21 August 1989. Snowploughs were obviously essential in the Scottish Highlands.

FOWEY DOCKS

D3497 (WDN) went new to Stratford (30A) in 1957 and stayed there until 1966 and therefore, as a youngster, I may well have seen this loco in Grays goods yard or later in Stratford shed; oh, for a time machine. Now at the end of its life in English China Clays, Fowey Docks, on 30 July 1989, I defy anybody to say they have seen an '08' in a more decrepit state than this. Built at Doncaster in 1957, after Stratford it went to Toton and then Colwick, withdrawal coming in 1968. It was then sold to English China Clays Ports, Fowey, but it is not believed to have worked for them. Somebody has painted '3497'; they have also painted 'RIP'. Yes indeed, 'Rest in Peace'. Or, should it be 'Rust in Pieces'?

Out of use and partly cannibalised by 1978, it was being broken up in 1981 but then work ceased and it remained heavily stripped and in a derelict state until 1989 when some spares were recovered for D3476; the remains were broken up on site in 1990. This date is 30 July 1989.

HOO JUNCTION STAFF HALT

Left: Darlington-built **09005 (SU)** (D3669) would soon be reallocated to Knottingley where I photographed it in a different livery in 1993, never realising at the time that I had also photographed it on a freezing winter morning at Hoo Junction two years previously. New to Norwood Junction in 1959, apart from Knottingley, future allocations were Brighton, Selhurst, Thornaby, Immingham and Toton. Coincidentally, 09005 ended its life in 'tactical store' here at Hoo Junction in 2009, 18 years after this image dated 19 February 1991 was taken. It was cut up by C. F. Booth of Rotherham in 2011.

Right: Brush-built Class 47/0 **47002 (TI)** *Sea Eagle* (D1522) passes through Hoo Junction Staff Halt with a mixed freight on 19 February 1991. New to Finsbury Park in 1963, I had seen it in Kings Cross station and Finsbury Park TMD in B.R. two-tone green on many occasions. It gained this unofficial name in 1989 at its current allocation, Sheffield (Tinsley). Future allocations were Stratford, Healey Mills, Gateshead, Crewe Diesel, Bescot, Cardiff Canton and Haymarket (Edinburgh). It would soon be withdrawn, June 1991, and it remained in store until 1994 when it was broken up by Booth Roe Metals, Rotherham. The station comprised two very short staggered platforms.

Deep and crisp and even. Class 73/1 **73133 (SL)** *The Bluebell Railway* (E6040) went new to Stewarts Lane in 1966. In 1990, it inherited its name from 73004 and carried it until 2004; the circular plate seen here above the name which reads 'Bluebell Railway' and 'Floreat Vapor'. The records I researched indicated that 73004 did not carry this circular plate but I photographed it at Woking station and I can assure you that it did. Further allocations were Eastleigh and Hither Green until, in 2003, it suffered an electrical fire in one of the cabs. After storage at Old Oak Common, it was restored and is preserved and still operational. Passing Hoo Junction Staff Halt in Civil Engineers 'Dutch' livery on a cold and frosty morning, this image is dated 19 February 1991.

LLANELLY STABLING POINT

Brush-built Class 47/3 **47315 (CD)** (D1796) and my Mark 3 Cortina HYJ 440N 'on the grid' at Llanelly TMD on 27 July 1980 (as detailed on page 155 of Stratford Depot Locomotives in The Eighties and Nineties, this car would be stolen from London's West End two months later on 22 September 1980). Although not realising it at the time, I must have seen 47315 many times on 'my patch' when it was D1796 and shedded at Stratford between 1968 and 1970. However, as can be seen in the inset, I had also seen it on 26 September 1971 at Crewe Diesel Depot, awaiting Works attention. New to Sheffield (Tinsley) in 1965, future allocations were Stratford, Thornaby, Toton, Crewe Diesel, Bescot, Cardiff Canton and Old Oak Common, where it was named Templecombe in 1993. The nameplates were removed at Toton in 1996, after which it spent time at Immingham, being taken out of use in 1999 and moved to Springs Branch CRDC where it was scrapped in 2000. There were four locos on shed at Llanelly on 27 July 1980: 08359, 08578 (a 'cop'), 37697 and 47315.

SEVERN TUNNEL JUNCTION TMD

On 27 August 1987, English Electric-built Class 37/9 **37905** *Vulcan Enterprise* **(CF)** (37136) (D6836) looks magnificent in the evening light, wearing Railfreight large logo grey livery. New to Cardiff Canton in 1963, future allocations were Landore (Swansea), York, Healey Mills, Sheffield (Tinsley), Gateshead then back to Canton in 1986, followed by the usual storage. It is now run by UK Rail Leasing and is operational in B.R. green as D6836; I'm happy about the colour but would prefer the running number to be in the correct font.

INVERNESS TMD AND STATION

Right: On 21 August 1989, Brush Class 47/4 (47776) **47578 (ED)** *The Royal Society of Edinburgh* (47181) (D1776) sits in Inverness Depot carrying standard large-logo Rail Blue livery. Having been allocated north of the border for nearly ten years it is no doubt a proud Scotsman although, unfortunately, it looked anything but when I photographed it two years later, as can be seen in the Eastfield (Glasgow) section of this book where the final chapter of its life is documented. New to Sheffield (Tinsley) in 1964, future allocations would be Finsbury Park, Stratford, Leeds (Holbeck) and Immingham, arriving at Eastfield in 1980 and gaining the above name at Edinburgh Waverley station in 1985. Although withdrawn in 1991, it was later reinstated.

Left: The building providing the backdrop is what it says on the tin… sorry… on the side of the Depot, 'ScotRail Inverness Depot'. Horwich-built Class **08 08754 (IS)** (D3922) went new to Corkerhill (Glasgow), further allocations being Leith Central, Polmadie (Glasgow), Eastfield (Glasgow), Dunfermline Townhill and Inverness. It is still operational for Rail Management Services. This image is dated 21 August 1989.

When I photographed Crewe-built Class 47/4 **47635 (IS) *Jimmy Milne*** (47029) (D1606) in standard large logo Rail Blue outside Inverness Depot on 21 August 1989, little could I have known that, 30 years later, it would be preserved virtually down the road to me. While photographing it, I was probably confusing Jimmy Milne with Jimmy Shand, the bandleader, not realising until now that, according to Google, the former was either a trade unionist or a footballer; take your pick. Going new to Landore (Swansea) in 1964, further allocations were Cardiff Canton, Old Oak Common, Bristol (Bath Road), Laira (Plymouth), Bescot and Eastfield (Glasgow) where it gained the above name in 1986. Allocations at Inverness, Sheffield (Tinsley) and Crewe Diesel followed and soon Jimmy Milne 'came out' as The Lass O' Ballochmyle. These nameplates were removed at Old Oak in 2004 and 47635 was sold to SECO Rail (Colas) in 2007. It is now preserved quite close to me at the Epping and Ongar Railway.

This is an interesting story; it is generally accepted that, in 1978, Tinsley was the first depot to experiment with unofficial names. However, in 1977, when the locomotive in this photograph had been allocated to Stratford, it caused a stir by appearing at Liverpool Street station carrying totally unofficial but professional-looking Great Eastern nameplates. The powers that be immediately ordered Stratford to remove them, but was this the catalyst for Tinsley to begin the avalanche of painted names that we are all aware of? On 21 August 1989, Crewe-built Class 47/4 **47460 (IS)** (D1580) had been an Inverness loco for six years and, apart from Inverness and Stratford, its tour of duty included Gateshead, Haymarket (Edinburgh), Immingham, Leeds (Holbeck), York, Eastfield (Glasgow) and Crewe Diesel. It gained the unofficial name of Triton in 1991, was withdrawn in 1992 and was cut up by Booth Roe Metals, Rotherham, in 1994.

Despite displaying an Inverness shed sticker from a previous allocation here, Crewe-built Class 47/4 **47643 (ED)** (47269) (D1970) is an Eastfield (Glasgow) engine. I note that, apart from photographing this loco two days earlier in Edinburgh Waverley station, I had not seen it since photographing it as 47269 at Motherwell shed in 1974. This image is dated 21 August 1989. Going new to Haymarket (Edinburgh) in 1965, further allocations were Gateshead, Eastfield (Glasgow) and Inverness. Withdrawn in 1991, 47643 is preserved at Bo'ness & Kinneil Railway. Complete with snow ploughs, it carries yet another variation of Executive or Main Line InterCity livery, as compared, for example, with 47551 in the Edinburgh Waverley station section of this book.

KINGS CROSS STATION

General view of Kings Cross station on 12 June 1992.

Left: Derby-built Class 08 **08407 (SF)** (D3522) sits in the locomotive bay at Kings Cross station on 15 March, 1991. Going new to Newton Abbot in 1958, it later worked out of Laira (Plymouth), Exeter, Truro, March, Colchester, Norwich (Crown Point) and Stratford, where it was apparently unofficially named Penguin although it shows no trace of that here; anyway, what would the cockney sparrow have made of it? It was used for work on the Channel Tunnel, Cheriton, but sustained motion damage and was withdrawn in 1993. It was scrapped that year by Gwent Demolition.

Right: Crewe-built Class 47/7 **47778 Irresistible** (47606) (47842) (47606) (47081) (D1666) displaying parcels livery sits in the same place on 3 May 1997. This photograph was taken two days after the General Election of 1 May 1997. The Conservative billboard clearly doesn't want Labour to do something; a modicum of basic research shows the full slogan is 'Don't Let Labour Blow It' but, in a landslide under Tony Blair, Labour ended their 18-year spell in opposition. 47778 went new to Landore (Swansea) in 1965 and was immediately named Odin. Later allocations were Cardiff Canton, Bescot, Old Oak Common, Bristol (Bath Road) and Crewe Diesel where it gained its current name in 1993. Transferred to Toton, these nameplates were removed and it then became Duke of Edinburgh's Award. Withdrawn in 2004, it was cut up the same year.

Left: A Power Car receiving power! On 18 January 1989, HST (High Speed Train) Class 43 Power Car **43064 (NL)** *City of York* is 'plugged in' in order that end-of-journey essentials that require electricity, such as lighting, cleaning, heating and air-conditioning, may take place when the Power Car's engines are turned off; only when it is moving are its batteries charging. This system is called 'shore supply', in the same way that cables run from the dockside to a ship in a cruise port. Regarding the livery; it looks similar to the original Power Car InterCity livery of yellow and blue but is, in fact, the later combination of yellow lower body and dark graphite upper section, separated by a white band; the passenger coach colours continue part-way into the Power Car. 43064 went new to Heaton (Newcastle) HST Depot in 1977. It also worked out of Craigentinny (Edinburgh) before this transfer to Neville Hill (Leeds). In 2008, it was renamed 125 Group and is still operational.

Right: As can be seen in this photograph taken over two years later on 15 March 1991, the Power Cars have now been given the same livery as the passenger coaches. In shot are **43110 (NL)** *Darlington*, **43088 (EC)** *X111 Commonwealth Games, Scotland 1986* and **43071 (EC)**. All are still operational. In 1979, 43110 derailed south of Northallerton due to low gearbox oil which caused the pinion to fail and lock the leading wheels of the train in place. With the rear power car pushing the train, the locked wheels skidded, wearing a groove which developed false flanges on their outsides, one of which struck the points, buckling the rail.

Brush-built Class 47/4 **47634 (SF)** *Holbeck* (47158) (D1751) in B.R. parcels livery actually on a parcels train. It went new to Landore (Swansea) in 1964, subsequent allocations being Bristol (Bath Road), Cardiff Canton, Old Oak Common, Immingham and Stratford where it was named Henry Ford. It was later allocated to Bristol (Bath Road) and Crewe Diesel where it was given the above name at Coalville Open Day 26 May 1991 with a plate below the name stating This locomotive was named at the last Coalville open day in recognition of the funds raised for local charities. I had photographed the original Holbeck (47425) in Liverpool Street station in May 1988 and 47634 will always be Henry Ford to me, as I saw it and travelled behind it when it was a Stratford loco. I note that on 12 July 1981, it pulled me on the 21.47 Shenfield to Liverpool Street as 47158 when I was on a 23.00 hours shift at the casino; it had only been named five days previously. This image is dated 12 June 1992.

Built at BREL Crewe Works in 1986, **89001 (BN)** entered regular passenger service between London Kings Cross and Peterborough in July 1988 after extensive testing by Crewe Electric depot on the West Coast main line and Hornsey and Bounds Green on the East Coast main line. The only Class 89 built, it was named Avocet in 1989, the nameplates being removed upon withdrawal in July 1992. After reinstatement at Bounds Green in 1997, it was given the nickname 'Aardvark' by railfans although it was alternatively known as 'The Badger' because of its slanted ends. Constantly beset with problems, I felt lucky to have caught it in revenue service on 3 May 1997 in the GNER blue and orange livery. It is now preserved.

Another view of this iconic locomotive.

Left: Crewe-built Class 90 **90022 (CE)** *Freightconnection* (90222) (90022) was the 'fourth musketeer', missing at Crewe Basford Hall when the three 'foreign' 90s were photographed (see earlier in this book). Going new to Willesden in 1984, it was later reallocated to Crewe Electric; it is currently in long-term store at Crewe IEMD and, after being left in the open for many years, it is unclear if it will return to traffic. This image is dated 8 June 1994.

Right: (47472) **97472** (47472) (D1600) was converted from a Class 47 locomotive in 1989 and taken into Departmental Service for the purpose of hauling test-trains throughout the country. Renumbered back to 47472 the same year it was allocated to Old Oak Common but suffered serious fire damage when in collision with 47533 at Reading in 1991. Towed back to its home shed, it was considered uneconomic to repair and was withdrawn, being broken up on site in 1997 by M. R. J. Phillips of Llannelly. Built at Crewe in 1964, it went new to Landore (Swansea) in 1964 and stayed on the Western Region until transferred to Eastfield in 1976 and then Inverness two years later. In 1983 it made the journey south to Stratford, a shed it called home for two years before moving on to Gateshead, Bescot and Crewe Diesel. This image of 97472 in standard Rail Blue livery is dated 15 March 1989. I had photographed this loco as 47472 at Liverpool Street station two years previously.

Left: (91118) **91018 (BN)** has always been allocated to Bounds Green. It would later carry two names, Robert Louis Stevenson and Bradford Film Festival (obviously not at the same time).

(91131) **91031 (BN)** *Sir Henry Royce* also had a small, circular plaque above the name which read ***Sir Henry Royce Memorial Foundation***. It, too, was only ever allocated to Bounds Green but was later named County of Northumberland. To commemorate the new speed record for a passenger train in 1995, it would later carry a plaque in the form of a laurel wreath upon which was inscribed: '91031 On 2nd June 1995 this locomotive powered the U.K.'s fastest passenger train which attained a speed of 154mph' (although back in 1989, 91010 had gone faster on a test train).

(91129) **91029 (BN)** *Queen Elizabeth II* was also only ever a Bounds Green loco. For some reason, in 1999, the nameplates were removed and reapplied with 'stick-on' letters.

They were all built at Crewe in 1990/1991; the date is 12 June 1992 and all three are still operational.

Right: Crewe-built Class 90 **90025 (CE)** was taken into stock at Willesden in 1989, later moving to Crewe Electric; it is currently in long-term store at Crewe IEMD and, after being left in the open for many years, it is unclear if it will return to traffic. This image is dated 22 April 1993.

Class 60 **60068 (TO)** *Charles Darwin* waits to haul Michael and me to Toton for the Open Day of 30 August 1998. It went new to Toton in 1991 and now has been shedded back there; in between it had been allocated to Immingham and Thornaby. Train enthusiasts come in all shapes and sizes and, it seems, apparel; I am trying very hard not to start humming, 'Donald, where's yer troosers'. Unfortunately, whilst hauling the 6H31 Margam to Llanwern in 2009, 60068 suffered serious engine and bodyside damage and is currently stored.

STEVENAGE STATION

Above: A different Class 60 was provided for the return journey and Stevenage was one of the stops on the way. I had journeyed out from and back to Kings Cross behind many Classes of steam and diesel, but never a '60'. Surely, therefore, two Class 60s in Kings Cross on the same day was a one-off, as I doubt it has ever happened before or since. Now in Mainline livery, **60044 (TO)** *Ailsa Craig* had gone new to Toton in 1991 and is now back there after a brief stint at Thornaby. It would have another spell at Thornaby in 2000 but went back to Toton again in 2012 for storage prior to heavy repair; it was later renamed Dowlow and is still operational. This image is dated 30 August 1998.

Left: My first sighting of 60044 was at Knottingley on 5 September 1993.

LAIRA (PLYMOUTH) TMD

Right: This image of **08641 (LA)** (D3808) on 30 July 1989 looks more like a painting than a photograph. Built at Horwich in 1959, it was always a Western Region loco, albeit well-travelled. Allocations were St. Philips Marsh (Bristol), Cardiff Canton, Newport (Ebbw Junction), Cardiff (Cathays), Landore (Swansea), Newton Abbott, Penzance (Long Rock), coming to Laira (Plymouth) in 1982. It was briefly unofficially named Dartmoor before eventually being sold to Great Western Trains; it was still operational in 2018.

Left: Built at Horwich in 1959, **08644 (PZ)** *Debbie* (D3811) was always a Western Region loco, allocations being Cardiff Canton, Newport (Ebbw Junction), Bristol (Bath Road), Penzance (Long Rock) and Laira (Plymouth), its last shed in 1988; it is not clear when it gained this unofficial name, nor when it lost it, but in 1991 it was given a different unofficial name, Ponsandane. Sold to Great Western Trains, it was still operational in 2018. It is seen here at Penzance station on 11 August 1986 (included here to keep all of the personalised Laira 08s together). A contrast here between a 'real' name and a 'home-made' job; I know which one I prefer. And for anybody wondering which Class 50 was named Furious, it was 50034. Like Count Arthur Strong, it could truthfully say, "I'm absolutely foorious!"

Left: Built at Derby in 1960, **08801 (LA)** *Plymouth* (D3969) was first allocated to St. Blazey, later allocations being Tyseley, Sheffield (Tinsley), Newport (Ebbw Junction), Cardiff Canton, arriving at Laira in 1988 where it was given this unofficial name the following year. In 1990, 08801 was working a rake of parcels vans from Long Rock to Penzance when it collided with 47538 which was leaving with the Penzance to Glasgow van train. After repairs were completed at Laira, it moved to Cardiff Canton and normally worked at Allied Steel and Wire. Withdrawn in 2000, it was cut up that year. This image is dated 30 July 1989.

Above: Built at Darlington in 1962, **08953 (LA)** (D4183) was first allocated to Cardiff Canton, later allocations being Newport (Ebbw Junction), Bristol (Bath Road), Newton Abbott, arriving at Laira in 1981 where it was unofficially named Plymouth. Presumably, it lost this name to 08801 in 1989 prior to 08953 being reallocated to Cardiff Canton, then Eastleigh and Immingham. In 2010 it was moved by road from Doncaster to European Metal Recycling, Attercliffe, Sheffield, and was finally scrapped in 2012. This image is dated 30 July 1989. The lucky mascot seems to be somewhat at odds with the smashed cab window. Although not the recipient of an unofficial name, a slight deviation from the norm has been achieved; Laira staff have personalised it to some degree by reproducing the running number and shedcode on the buffer beam. Unusually, it carries a note of date and Depot of last repaint next to the cab door ('repainted 21.4.86 LA').

The Western Region clearly received a batch of 08s from Horwich in 1959, and here is another. **08645 (LA)** *Friary* (D3812) went first to Cardiff Canton, followed by Newport (Ebbw Junction), Landore (Swansea), Margam, arriving at Laira (Plymouth) in 1983; it is not clear when it gained this unofficial name but, soon after this image was taken, it was repainted in 'B.R. General all-over mid-grey' offset by black window surrounds and a dark grey roof; the name disappeared from the fuel tank side, replaced by a white running number. Always a Western Region loco, it is now owned by Great Western Trains and later carried another unofficial name, St. Piran. This image is dated 30 July 1989.

It now transpires, mildly disappointingly, that I had seen this loco many times between 1968 and 1977 when it carried a different running number and was shedded at Stratford. But, on 30 July 1989, seeing it at Laira in Railfreight Metals livery with a different running number and now with an attractive Cornish name, how could I have possibly known? English Electric-built Class 37/5 **37671 (LA)** *Tre Pol and Pen* (37247) (D6947) went new to Cardiff Canton in 1964, future allocations being Healey Mills, Stratford, Eastfield (Glasgow), and Laira (Plymouth). In January 1988, 37671 and 37672 were hauling a freight train that was diverted into a siding at Tavistock Junction, Devon, due to a pointsman's error. The train collided with a wagon, pushed it through the buffers and was derailed. Just over a year later, the repaired 37671 collided with 37674 at St. Blazey; both locos were repaired and returned to service. Allocated to 'Special Projects', it went to France in 1999, returning after just over a year. Long-term storage at Tyne Yard followed until it was taken by road to European Metal Recycling, Attercliffe, Sheffield, and was broken up in 2011. The phrase 'Tre Pol and Pen' is used to describe people from, or places in, Cornwall. The full rhyming couplet is: 'By Tre Pol and Pen shall ye know all Cornishmen'.

Class 50 **50018 (LA)** *Resolution* (D418) sits under Laira's hoist in Network SouthEast livery on 30 July 1989. New in 1968, it worked out of Crewe Diesel, Bristol (Bath Road) and Laira (Plymouth), withdrawal coming in 1991. It was scrapped in 1993 at M. C. Metal Processing, Springburn Works, Glasgow.

New to Cardiff Canton in 1963, English Electric-built Class 37/5 **37673 (LA)** (37132) (D6832) in Railfreight Distribution livery had been at Laira for a couple of years on 30 July 1989. Between times, it had resided at Gateshead, Healey Mills, Thornaby, Stratford, Sheffield (Tinsley) and Immingham. Later, it went back to Canton where it began life, before allocations to Bescot, Toton and Crewe Diesel as a Sandite loco. After storage at Thornaby, it was taken by road to European Metal Recycling, Kingsbury, in 2008 and was broken up that year. When new to Laira, it worked into London and I have a wonderful photograph of it in pristine condition in Liverpool Street station, being used by Stratford on an empty stock working.

On 30 July 1989, English Electric Class 50 **50040 (LA)** *Centurion* (D442) has only a few more months of service left, being withdrawn in November 1989. Going new to Crewe Diesel in 1968, it later worked out of Bristol (Bath Road), Old Oak Common and Laira (Plymouth). Interestingly, this loco carried three names in its time; one set inappropriately, one 'real' set and one set in error. It was first named Leviathan in 1978 but the plates were removed in 1987 because the aircraft carrier with this name was never completed and lay unfinished in Plymouth Dockyard from 1945 to 1968, when it was broken up; it was deemed inappropriate that a loco be named after a ship that never wholly existed. In July 1987 it was given the name it carries in this image but, when in Doncaster Works later that year along with 50033 Glorious, it had the latter's nameplates fitted by mistake. Back at Doncaster after a test run to Peterborough, the mistake was noticed and rectified. 50040 was scrapped at Sims Metal Management, Halesowen, in 2008.

Crewe-built Class 47/9 **47971 (CD)** *Robin Hood* (97480) (47480) (D1616) went new to Toton in 1964 and spent the next ten years based on various divisions of the Midland Region, until reallocation to Crewe Diesel. Back to Toton in 1979, 47971 received the above name at Nottingham station that year. After spells at Carlisle (Kingmoor) and Bescot, it returned to Crewe Diesel in 1994. Withdrawn in 2000, the following year it was moved by road to European Metal Recycling, Kingsbury, and broken up by Harry Needle Railroad Company. This image is dated 30 July 1989. I have a note that I saw it as 47480 Robin Hood in Stratford carriage sidings on 20 July 1988.

Left: By 1992, **50015 (LA)** (D415) was one of just eight Class 50s remaining in service, the others being 50007/008/029/030/033/046/050; it once carried the name Valiant and was the only '50' to be repainted in 'Dutch' Civil-Engineers grey and yellow livery. Seen at Laira on 7 March 1992, it was withdrawn the following June; initially purchased from B.R. by Pete Waterman, it was later transferred to the 'Manchester Class 50 Group' for preservation and is now owned by the Bury Valiant Group. Allocated to Crewe Diesel from new in 1968, 50015 was reallocated to Laira in 1974.

Right: English Electric Class 50 **50037 (LA)** *Illustrious* (D437) built in 1968 moved from Crewe Diesel to Bristol (Bath Road) in 1974, later working out of Laira and Old Oak Common; arriving at Laira for a second time in 1989, withdrawal coming in 1991. In August 1966, DP2 derailed in Edinburgh Waverley station and the following year it was involved in a serious accident at Thirsk, colliding with a cement train. After storage at York shed for two months where it remained sheeted over it was withdrawn from service and moved to the Vulcan Foundry where it was dismantled, its reusable parts being provided to the Class 50 pool of spares. DP2's engine initially went to D417 (50017) but ended its working days in D437 (50037). So, in this image dated 30 July 1989, DP2's engine is inside the body of this Network SouthEast liveried loco which was scrapped by M. C. Metal Processing, Springburn Works, Glasgow, in 1992.

PLYMOUTH STATION

It is difficult to comprehend the number of lives lived by almost every main-line diesel, and 1967 Brush-built Class 47/7 (57303) **47705 (OC)** (47554) (D1957) sitting at Plymouth station is a prime example. It was allocated to Crewe Diesel, Bescot and Haymarket (where it was named Lothian) in 1979, running in ScotRail livery. After that came Eastfield (Glasgow), who removed the nameplates in 1989 when the loco was transferred to Old Oak Common; Eastleigh came next and then Crewe Diesel for a second time. Sold to Waterman Railways in 1994, it was named Guy Fawkes at the Crewe Rail Fair at Basford Hall Yard in 1995 in a ceremony which Michael and I witnessed (see inset). Conversion to a Class 57 came in 2003 when 47705 was rebuilt as a Virgin-liveried Thunderbird, 57303 Alan Tracy. Later, when a Direct Line loco, it was given yet a fourth name, Pride of Carlisle. This image is dated 30 July 1989.

PENZANCE STATION

A couple of years after this image was taken at Penzance station on 11 August 1986 with my son standing next to me, I would be photographing him standing next to Class 47/4 **47522 (GD)** (D1105) in Stratford Depot on 22 October 1988, an image that would later appear in Stratford Depot Locomotives in The Eighties and Nineties published by Mortons Books. By then, 47522 would be allocated to Crewe Diesel, named Doncaster Enterprise, and be uniquely liveried in experimental LNER-style apple-green. Here though, at Penzance station on 11 August 1986, it is a Gateshead loco as the shed sticker and crest indicate and it carries standard B.R. Rail Blue livery with full yellow warning ends. I also photographed it at Liverpool Street station in July 1987 while it was still a Gateshead engine.

I had witnessed this locomotive being built on a visit to Crewe Works on 3 April 1966 and I have notes of seeing it many times since. In 1982 it was badly damaged when it hit a tractor on a farm crossing near Forteviot at 55mph (from 90mph) when hauling the 13.35 Glasgow Queen Street to Aberdeen. It ripped up 300 yards of track and plunged 35ft down an embankment. Extensively damaged (every bit of bodywork above the base frame was written off) and needing one completely new cab, it was repaired at Crewe Works, the cost being in excess of £200,000. 47522 was involved in another accident near Dover in 1990 and received a second new cab, this time from 47645, which was stored at Doncaster Works with serious fire damage due to an accident. After this second rebuild, 47522 was out-shopped in red and graphite parcels sector livery and I note that I saw it as such at Kings Cross station on 16 April 1992. Withdrawn in 1999, 47522 was scrapped in 2000 by EWS at Springs Branch (Wigan) CRDC. Modellers, please note piles of sleepers on platform, not to mention a motorcycle.

It looks a bit misty off the Penzance coast; I can't spot any pirates but clearly visible is Class 50 **50045 (LA)** *Achilles* (D445), built by English Electric in 1968 and carrying large logo blue livery; it is about to depart Penzance station with a Travelling Post Office train (TPO) for London Paddington, the Class 50 being the loco of choice for this duty. 50045 was named in 1978 after transfer from Crewe Diesel to Laira (Plymouth), these being its only allocations. This loco was withdrawn from service in 1990 and, after ten years of storage and dumping, was broken up by Booth Roe Metals, Rotherham, in 2000. The date is 11 August 1986 and, purely out of interest, in the final episode of the TV detective drama Wycliffe entitled 'Land's End' and broadcast in 1998, Detective Superintendent Charles Wycliffe (Jack Shepherd) arrives here at Penzance station on Class 153 sprinter number 153382, uses a station telephone box and gets into a taxi standing next to HST number 43127. Also, interestingly, another HST, 43097, is briefly glimpsed at a fictitious station (filmed at St. Erth) in a 1996 episode entitled, 'Dead on Arrival'.

LANDORE (SWANSEA) TMD

Crewe-built Class 47 (47741) **47597 (CD)** *Resilient* (47026) (D1597) began life at Cardiff Canton in 1964 and was reallocated to Landore (Swansea) the same year. It had therefore probably stood on this very spot many times in BR two-tone green however, on 7 March 1992, it has gained red and graphite parcels sector livery and is now allocated to Crewe Diesel. During its working life it had also worked out of Bristol (Bath Road), Bescot, Laira (Plymouth), Old Oak Common, Eastfield (Glasgow) and Toton. In 1999, whilst hauling the Swansea to Willesden Railnet mail train, it became yet another Class 47 to catch fire. Following long-term storage at Toton, it was broken up by European Metal Recycling, Kingsbury, in 2008. This image is dated 7 March 1992.

Crewe-built Class 08 **08993 (LE)** *Ashburnham* (08592) (D3759) went new to Danygraig (Swansea) in 1959, followed by allocations to Old Oak Common, Southall, Cardiff Canton and Reading before coming to Landore (Swansea) in 1980. In 1985, it gained its current number when it was modified from the standard class by being given headlights and cut-down bodywork in which the overall height was reduced to 11ft 10in (3.61m), for use on the Burry Port and Gwendraeth Valley Railway up to Cwm Mawr. However, it was reported to be stored much of the time until moving to Cardiff Canton in 1994. This image is dated 7 March 1992 and 08993 is now preserved on the Keighley and Worth Valley Railway.

LEICESTER TMD

One way into Leicester TMD was through a fence and a few bushes, up this ramp inside what appeared to be an abandoned goods shed and then down the other side. Whilst the 'new' Class 60s were being built at the Brush Works, I always visited Leicester TMD en route to Loughborough and the same three shunters were always in the same spot as pictured below, never moving; the one in the middle was 08473. The two pictured began their working lives at Laira so it is strange that they should end those working lives growing derelict together here.

Left: Swindon-built Class 03 **03197 (WDN)** (D2197) went first to Laira (Plymouth) in 1961, later allocations being Taunton, Feltham, Bournemouth, Eastleigh, Slade Green (Dartford), Norwich (Thorpe), and Norwich (Crown Point). Withdrawn in 1987, it was moved to Leicester where it lay in a derelict condition with parts being used for the construction of the new diesel built for the Vale of Rheidol Railway. In 2001 it was moved to Shackerstone, Battlefield Line, later going to Mangapps Farm Railway Museum in 2010. This image is dated 2 June 1991.

Right: Derby-built Class 08 **08399 (CD)** (D3514) went first to Laira (Plymouth) in 1958, later allocations being Derby, Toton, Longsight (Manchester), Newton Heath (Manchester), Derby (where it was unofficially named Great Central) and Crewe Diesel. In 1993, 08399 was broken up by M. C. Metal Processing, Springburn Works, Glasgow. This image is dated 2 June 1991.

After making it through the bushes and turning towards the main part of the Depot, the intrepid spotter was always treated to what I called 'The Leicester line-up'; this was the sight greeting me on 24 May 1992. Front locos only, left to right: **60010 (TO)** *Pumlumon / Plynlimon*, **60011 (TO)** *Cader Idris*, **56045 (TO)**, **56063 (TO)** *Bardon Hill*, (47314) (47387) **47314 (TI)** *Transmark* (D1795), **60012 (TO)** *Glyder Fawr*.

Crewe-built **47972 (CD)** (97545) (47545) (D1646) went new to Cardiff Canton in 1964, future allocations being Landore (Swansea), Bristol (Bath Road), Bescot, and Crewe Diesel where it was named The Royal Army Ordnance Corps. It went on to work out of Immingham and was broken up by C.F. Booth of Rotherham in 2010. It carries large logo Rail Blue livery and this image is dated 24 May 1992.

Left: **56032 (CF)** *Sir De Morgannwg / County of South Glamorgan* (names reversed on other side) was taken into stock at Toton in 1977, later allocations being Sheffield (Tinsley), Healey Mills, Cardiff Canton, Thornaby and Immingham. It is currently with Boden Rail Engineering. This image is dated 10 March 1991.

Right: As can be seen, the position of the sun was not conducive to obtaining a good shot from this angle but I had to settle for it. **45149 (WDN)** *Phaeton* (D135) went new to Derby in 1961, future allocations being Kentish Town, Toton and Sheffield (Tinsley), where it gained this unofficial name. Withdrawn in 1987, it was stored at Cricklewood in an increasingly derelict condition before being moved to Leicester in 1991. Sold to Pete Waterman for preservation in 1992, it currently resides at the Gloucestershire-Warwickshire Railway, looking rather different to this image dated 24 May 1992.

60059 (NYA) *Samuel Plimsoll* had not yet been allocated to a Depot when this image was taken on 2 June 1991. The next month, Thornaby was the lucky recipient, later allocations being Toton, Immingham and Cardiff Canton. It kept its original nameplates for a shorter period of time than any other Class 60, becoming Swinden Dalesman in 1995; it is still operational.

Taken into stock at Toton in 1992, **60083 (TO)** *Shining Tor*, also worked out of Thornaby, changing its name to Mountsorrel along the way. This image is dated 16 April 1995; along with most of the class, it remains stored (or rather, abandoned) in very faded EWS colours in Toton yard (sorry, graveyard). Amazingly, a few of these sad-looking locos still retain their nameplates; in an ever-increasing state of dilapidation, what on earth is going to happen to them all? One thing is certain, 60083 now looks nothing like it does in this image.

MOTHERWELL TMD

Left: It is 20 August 1989 and **20203 (WDN)** (D8303) had been withdrawn the previous year; it would remain stored at Motherwell until 1992 when it would be moved to M. C. Metal Processing, Springburn Works, Glasgow. Going new to York in 1967, future allocations were Sheffield (Tinsley), Haymarket (Edinburgh), Immingham and Eastfield (Glasgow), for scrapping.

Right: Robert Stephenson & Hawthorns-built Class 37/0 **37077 (TI)** (D6777) was a loco very familiar to me as it had been shedded at Stratford a couple of times and, lurking in the background, were a couple of current Stratford charges, complete with cockney sparrows. I wonder what they made of 'the sparrer' north of the border when he first appeared? Going new to Thornaby in 1962, future allocations were Healey Mills, March, Gateshead and Sheffield (Tinsley). On its fourth spell at Thornaby it was named British Steel Shelton but only kept the name for two years. Then came Sheffield (Tinsley) (who unofficially named it Hurricane), Motherwell, Cardiff Canton, Bristol (Bath Road), Eastleigh and Toton. Two spells in France followed until in 2009 it was cut up by C. F. Booth of Rotherham. This image is dated 20 August 1989. Note that no two fronts are alike.

Robert Stephenson & Hawthorns-built Class 37/0 (37088) **37323 (ML)** *Clydesdale* (37088) (D6788) went new to Gateshead, future allocations being York, Dairycoates (Hull), Sheffield (Tinsley), March, Eastfield and Motherwell, where it was given the above name. In 1999 it was taken by road to Springs Branch (Wigan) CRDC; then by road to Barrow Hill where it was stripped for spares and thence by road in 2002 to Booth Roe Metals, Rotherham, for scrapping. Triple time for Allelys! In all-over B.R. Rail Blue, this image is dated 20 August 1989.

English Electric-built Class 37/0 **37111 (ML)** *Glengarnock* (37326) (37111) (D6811) went new to Sheffield (Darnall) in 1963, future allocations being Sheffield (Tinsley), Stratford, March and Eastfield, where it was named Loch Eil Outward Bound. Next came Motherwell, who clearly thought Glengarnock was a better name. It later worked out of Immingham, Inverness and Bescot before withdrawal in 1998. It was scrapped by the Harry Needle Railroad Company in 2003. This image of 37111 carrying Railfreight Metals triple grey livery is dated 20 August 1989.

Brush-built Class 47/0 **47137 (WDN)** (D1729) went new to Landore (Swansea) in 1964, later allocations being Bristol (Bath Road), Cardiff Canton, Laira (Plymouth), Crewe Diesel, Carlisle (Kingmoor) and Eastfield (Glasgow). Withdrawn in 1987, it is seen here stored at Motherwell and was eventually broken up by M. C. Metal Processing, Springburn Works, in 1992. This image is dated 20 August 1989 so the faded B.R. Rail Blue paintwork has a way to go yet.

OLD OAK COMMON TMD

Old Oak Common has recently entered the public consciousness as the new proposed terminus for the High-Speed train (as opposed to Euston) but, prior to that, I would hazard a shrewd guess that, even if they had vaguely heard of it, most people would be hard pressed to locate it on a map. However, amongst the train enthusiast community, we had known of Old Oak for many years, bunking the shed through steam and then diesel days until, like most other TMDs, it took an overdose of self-importance pills and erected electronic gates and high metal fences. Apparently, a trainspotter trying to 'cop' his last Class 08 posed such an overwhelming threat that such extreme measures were deemed essential.

Left: All dressed up but definitely having somewhere to go; sitting in the 'new' three-road diesel refuelling shed in Executive InterCity livery, we can only speculate upon the possibility of Crewe-built Class 47/4 (47788) **47833 (BR)** (47608) (47262) (D1962) hauling Royal Claret coaches from Paddington to Windsor. New to Cardiff Canton in 1965, this loco later spent time at Crewe Diesel, Bescot, Old Oak Common, Bristol (Bath Road) and was named Captain Peter Manisty RN in 1993. The following year, it received front-end damage whilst propelling empty coaching stock into Bridlington station, but was repaired. In 2000, it was broken up at Springs Branch (Wigan) CRDC. This image is dated 5 November 1989.

Built at Crewe in 1964, (47722) **47558 (BR)** *Mayflower* (47027) (D1599) began life at Landore (Swansea), later moving to Bristol (Bath Road), Cardiff Canton, Old Oak Common, Laira (Plymouth) and back to Landore in 1982 where it was given the above name after the nameplates were taken by road from Laira (where it was originally intended to be named). These came off in 1988, after which it became Carlisle Currock. However, in British Rail's favourite game of musical nameplates, this name was removed in 1990 and it became Mayflower once more. But B.R. must have thought, why stop there? So, off came the nameplates again, only to be replaced with The Queen Mother in 1995; and yes, after much storage and reinstatement, these were removed in 2002. Transported by road from Toton's Training Compound to European Metal Recycling, Kingsbury, the loco was broken up soon after arrival. In early-style InterCity livery with yellow warning ends, this image of 47558 is dated 5 November 1989; to the left can be seen the roof of the 'new' three-road diesel refuelling shed while to the extreme right is the roof of the Pullman Shed.

Left: Although an Old Oak loco on 26 January 1985, Brush-built Class 31/1 **31135 (OC)** (D5553) spent most of its life on the Eastern Region at Ipswich, Norwich (Thorpe), Stratford and March. It also spent a small amount of time at Immingham, Sheffield (Tinsley), Bescot and Crewe Diesel until it was withdrawn from Toton in 1998 and scrapped by T. J. Thomson of Stockton in 2000. Neither me, nor friends with whom I have shared this image, have ever seen a Class 31 carrying what we think must be a livery unique to this loco; i.e. all-over B.R. Rail Blue but with the yellow warning front continued around the cab windows, like the Class 47 next to it.

Right: This locomotive was yet another Class 47 to catch fire, this time in Birmingham New Street station in 1999 whilst on a Wolverhampton to Plymouth service. It caused extensive fire damage to the station's overhead ducting and the loco was removed to Saltley for examination and repair. Brush-built Class 47/4 (47839) **47621 (BR)** *Royal County of Berkshire* (47136) (D1728) went new to Cardiff Canton in 1964, future allocations including Landore (Swansea), Bristol (Bath Road), Laira (Plymouth) and Old Oak Common. Seen here in yet another variation of InterCity livery, this image of 47621 is dated 7 May 1988. It was later to carry two more names, Pride of Saltley and Pegasus before being broken up on site at Eastleigh Works by Raxstar in 2013.

Brush-built Class 47/4 (47742) **47598 (OC)** (47182) (D1777) went new to Sheffield (Tinsley) in 1964, further allocations being Stratford, Leeds (Holbeck), Crewe Diesel, Toton, Bescot, Laira (Plymouth), Old Oak Common and back to Crewe Diesel in 1991, where it was named The Enterprising Scot in 1995. Withdrawn in 2001, after storage at Toton, it was taken by road to European Metal Recycling, Kingsbury, in 2007 and cut up that year. This image of 47598 inside 'the factory' in Network SouthEast livery is dated 5 November 1989.

Built by English Electric in 1968, **50025 (OC)** *Invincible* (D425) did not live up to its name because, on 6 August 1989 whilst hauling the 21.15 Oxford to Paddington passenger train, it hit a metal rail placed across the rails by vandals, tipped over on its side and slid along the track and onto the platform at West Ealing station, suffering massive damage. After two days, it was moved the short distance to the Waitrose Supermarket car park who, if they are anything like my local Tesco, have a three hour 'maximum stay'. It was later placed on bogies and slowly moved to Old Oak Common and withdrawn from service six days later. The two images seen here show it at Old Oak in pristine Network SouthEast livery on 7 May 1988 and after the derailment on 12 September 1989. The following month, staff from Vic Berry cut it into movable chunks of scrap and moved it to their Leicester yard. An image dated 20 January 1990 showing just the two remaining cabs may be seen later in this book in the Vic Berry scrapyard section.

Looking resplendent, Class 47/5 **47515 (OC)** *Night Mail* (D1961) is pictured in 'the factory', still an Old Oak loco on this date but it would be reallocated to Crewe Diesel later in the month. I photographed it in Kings Cross station about nine months later and it was certainly not in this condition. New in 1968, it worked on various divisions of the LMWL before allocations to Crewe Diesel, and Old Oak Common, gaining the above name in 1986. It was withdrawn from service in 1991 and cut up; in a version of Executive InterCity livery, this image of 47515 is dated 7 May 1988.

Left: This image taken about halfway along 'the factory' at Old Oak clearly shows the size of it. Brush-built Class 47/4 **47512 (OC)** (D1958) went new to LMWL in 1967, future allocations being Landore (Swansea), Old Oak Common, Gateshead and Crewe Diesel. Withdrawal in 1991 was followed by storage at Crewe diesel until it was cut up by Booth Roe Metals, Rotherham, in 1994. This image of 47512 in standard B.R. Rail Blue livery is dated 26 January 1985. Old Oak always had a good turnover of Class 08 shunters and **08634 (OC)** (D3801) in the background had recently come from Ebbw Junction (Newport). It is interesting to note that it was sold for preservation and the last report I can find is that it has been renumbered D3801 and stored at Barrow Hill Roundhouse Museum. Don't miss the raised '50' poking through the doors.

Right: Brush-built Class 47/4 **47501 (OC)** (D1944) went new to LMWL (London Midland Western Lines) in 1966, future allocations being Bescot, Crewe Diesel, Laira (Plymouth), Landore (Swansea) and Old Oak Common where it gained the name Craftsman in 1987. My image of it as such, repainted in 'parcels' livery, is depicted in Stratford Depot Locomotives in The Eighties and Nineties. The last allocation for 47501 was Bristol (Bath Road) after which it has undergone lengthy storage. The last report I can find is that it still remains in store for Direct Rail Services. This image of 47501 in standard B.R. Rail Blue livery is dated 26 January 1985. As can be seen, HSTs were also repaired in 'the factory'.

VIC BERRY SCRAPYARD, LEICESTER

Not the warmest of welcomes but I need not have worried as I cannot remember ever seeing a soul. Modellers take note; from the weathering and construction of the (girder) bridge to the cinders, clinker and moss between the tracks. Just under the girder bridge, though, without doubt the sight that greeted visitors to Vic Berry's yard certainly gave one a feeling of total incredulity; mountains of B.R. and Underground coaching stock mixed with many classes of diesel locomotives piled high in the sky plus scattered cabs and body shells of more diesel locos at ground level was mind-boggling.

Left: Seen here in all-over B.R. Rail Blue livery, Class 31/1 **31127 (WDN)** (D5545) was withdrawn in 1989 after suffering very extensive fire damage at Catcliffe, a light engine movement turning into an engine alight movement. Spending the first 13 years of its life at Ipswich, March, Stratford and Norwich, it later divided its time between Immingham, York, Holbeck (Leeds), Healey Mills, Bescot, Crewe Diesel, Stratford and Thornaby. After the fire, it spent a short time in storage at Tinsley before it was moved to the infamous Vic Berry scrapyard in Leicester and broken up in 1990. Photographed in Vic Berry's yard on 20 January 1990, the fire damage is apparent; the two scrawls on this end of the loco appear to contradict each other. 31127 also appears in sister publication, Stratford Depot Locomotives in The Eighties and Nineties.

Right: Built at Beyer Peacock Works, Gorton, in 1966, Class 25/3 **25907 (WDN)** (25297) (D7647) was scrapped by Vic Berry in 1989. It went new to Sheffield (Tinsley), future allocations being Staveley (Barrow Hill), Longsight (Manchester), Toton, Springs Branch (Wigan), Crewe Diesel and Carlisle (Kingmoor). Withdrawn in 1986, it was stored at Springs Branch (Wigan) until it was moved to Vic Berry's yard in 1987 and cut into pieces. This image is dated 20 January 1990. A very thoughtful person has daubed its last two numbers on the front. Don't miss the red London Transport cab-front looking totally out of place, top right in the stack of mostly Mark I coaches. I am reliably informed it is 1938 'P' Stock. Also, I hope I am wrong, but is that oil drum next to all those propane gas cylinders in use as a brazier? And people wonder how 'The Great Fire of Leicester' started!

Anyone call a cab? When these English Electric Class 20s were introduced in 1966/1967, I travelled with Michael in hired Minis to as far afield as Scotland to see the 'new D8000s' (as we called them), now being built with central headcode boxes. I never dreamed I would be travelling to see them in this sad state at the end of their lives. **20183** (D8183) and **20150** (D8150) had been predominately Midland Region locos, 20150 being withdrawn in 1987 and, due to an error, 20183 was found to be never officially withdrawn when it reached Vic Berry's yard. **20201** (D8301) had been a Scottish loco for the majority of its life and was withdrawn in 1988. All three were cut up in 1990, this image being dated 9 June 1990.

The only running number visible is that of Class 25 **25327** which had been withdrawn in 1984 and, after the obligatory storage, had arrived at Vic Berry's yard in June 1987. Three years later, on 20 January 1990, one cab is all that remains of it. By an amazing coincidence, at Carlisle (Kingmoor) on 28 May 1984, I had photographed my son, Kevin, with a Class 40 nose either side of him; and what cab and number is visible between them in the next line of locos? Yes, I'm sure you are way ahead of me.

Left: Brush-built Class 47/4 **47415 (WDN)** (D1514) went new to Finsbury Park in 1963, later allocations being Sheffield (Tinsley), York and Gateshead. Withdrawn in 1987, it spent three years in various places of storage including Tyneside Central Freight Depot and Whitemoor Yard until it arrived at Vic Berry's in 1990 and was cut up soon after. The date is 20 January 1990.

Right: This picture shows just some of the diverse stock that Vic Berry broke up in his yard. At first glance, GUV parcels vans can be spotted, cabs of a '25' a '50' and an '08' and what appears to be a Mark 1 EMU. The London Underground carriage is 1983 stock with single-leaf doors built by Metro-Cammell for the Jubilee Line. In the foreground, British Rail Class 303 number **303 057** was originally termed a 'blue train' when first introduced in 1960 for the electrification of the North Clyde and the Cathcart Circle lines in Strathclyde. The fleet had a lifespan of 42 years and 50 of the units were refurbished in 1984, ending their days painted in the orange and black livery of Strathclyde PTE.

Left: An almost identical C.V. to 47415. Brush-built Class 47/4 **47419 (WDN)** (D1518) went new to Finsbury Park in 1963, later allocations being Sheffield (Tinsley), York and Gateshead. Withdrawn in 1987 due to fire damage, it then spent three years in various places of storage including Tyneside Central Freight Depot and Whitemoor Yard before being moved with 47415 to Vic Berry's yard in January 1990 and was cut up that year.

Right: Fancy a nap? We were constantly surprised as to what would turn up next in Vic Berry's yard and here is a good example; an Inter-City Sleeper train. What more is there to be said?

Class 31/4 **31443 (WDN)** (31177) (D5598) began life at Hornsey in 1960, followed by allocations at Finsbury Park, Stratford, Immingham, March, Toton and Crewe Diesel. Withdrawal came in 1989 due to severe fire damage sustained near Tollington whilst hauling an HST Barrier Vehicle. After storage at Doncaster, 31443 was moved to Vic Berry's yard in 1990 and cut up. This image is dated 9 June 1990. On holiday as a schoolboy, I photographed 31443 (as D5598) in Walton-on-the-Naze goods yard when we were both nothing more than spring chickens (see inset).

'A Tale of Two Cabs' or 'Talking Heads'? If the latter, I wonder if they would be saying, "How on earth did we end up here?" For the unfortunate story of exactly how the two cabs of 50025 did end up here, please refer to the 'Old Oak Common' section of the book.

Left: Vic Berry was apparently alert to the fact that 'flame cuts' were collectable. On 20 January 1990, on display are 25199, 08606, 31189 and 47140.

Right: On 19 July 1991, the day after British Rail reclaimed the land, burnt carriages are piled up after 'The Great Fire of Leicester' which occurred on 10 March 1991; this was exactly three months prior to Vic Berry ceasing trading on 10 June 1991. The fire destroyed approximately 150 Mark 1 coaches stacked on top of each other and it took 85 firefighters to control the flames. Reports varied regarding the cause; Vic Berry had financial problems and arson was strongly suspected but combustion connected with the removal of asbestos was also mooted; after a large smoke-cloud containing asbestos particles blew across the city, the question was belatedly asked as to how such an operation was allowed to take place in the middle of a major city in the first place. A subsequent investigation relating to the fraudulent procurement of locos and stock was later dropped. After 26 years in the business and despite plans to begin another operation, Vic Berry died on 20 March 2015 aged 82.

CONTENTS

A1 (south of Wetherby) ... 43	Exeter St Davids station ... 114	Nottingham holding sidings ... 88
Ayr TMD ... 18-19	Fort William stabling point ... 130-131	Old Oak Common TMD ... 176-182
Barry TMD / station ... 52-53	Fowey Docks ... 132-133	Penzance station ... 152, 162-163
Basingstoke station ... 20-21	Gatwick station ... 54-61	Plymouth station ... 161
Birkenhead North TMD ... 109	Hoo Junction ... 58, 134-135	Reading TMD ... 114
Blyth Cambois (TMD / surrounds) ... 89-91	Immingham TMD ... 62-63	Severn Tunnel Junction TMD ... 137
Bounds Green TMD ... 14-15	Inverness (station / TMD) ... 138-141	Shenfield station ... 64-70, 75, 78
Camden Road station ... 71, 73	Kensington Olympia station ... 82-83	Shildon Museum ... 94
Cardiff Canton TMD ... 22-27	Kings Cross station ... 142-150	St. Blazey TMD ... 115-117
Carlisle (Citadel station / Upperby TMD) ... 110-113	Knottingley TMD ... 1-11, 13, 151	Stevenage station ... 151
Chelmsford (area) ... 13, 118-121	Laira [Plymouth] TMD ... 152-160	Stratford station and surrounds ... 15, 48, 50-51, 72
Clapham Junction station ... 84	Landore TMD ... 164-165	Thornaby ... 160
Crewe (area) ... 92-108, 136	Leicester TMD ... 166-171	Tilbury Dock ... 32
Doncaster (TMD / station / sorting sidings) ... 28-30	Liverpool Street station ... 72	Toton ... 46
Doncaster (Works) ... 31-35	Llanelli (Crewe Diesel) ... 136	Vic Berry's Yard ... 183-191 (Walton-on-the-Naze station 189)
Eastfield (Glasgow) TMD ... 34, 36-49	Loughborough ... 16-17	Waterloo station ... 85-87
Eastleigh ... 60	Mitre Bridge Junction ... 76, 80-81	Watford Junction station ... 74, 79,
Edinburgh Waverley station ... 122-123	Motherwell TMD ... 172-175	Willesden TMD ... 77
Euston station ... 124-129	National Power Ferrybridge ... 12	